NAVIGATION RULES

for
International and Inland Waters

edited by
David Burch

Paradise Cay Publications
Middletown, CA

Published by
Paradise Cay Publications
P.O. Box 20, Middletown, CA 95461

Manufactured in the United States of America

ISBN 0-939837-16-1

4 5 6 7 8 9 10

Preface

The goal of this publication is to provide mariners with a concise yet complete copy of the Inland and International Navigation Rules to be carried on board for quick reference or study.

Section 88.05 of Annex V to the Inland Rules requires that an *up-to-date* copy such as this one be carried on all vessels 12 meters (39 feet) or more in length at all times on Inland Waters. It is clearly prudent, however, to have these rules available on all vessels, in all navigable waters. The COLREGS Demarcation Lines, included at the end of this book, specify the official boundaries between Inland and International Waters as regards the application of the Rules.

We hope that this publication will fulfill the clear need for a readily available, up-to-date text to meet the "on-board requirement" of the Rules. We don't hesitate to stress the obvious, however, that having the Rules on board is not adequate in itself for a safe and proper application of the Rules. This requires study. Having a copy of the Rules in hand is the first step.

Study of the Rules is a rewarding pastime, practical and captivating. They constitute a remarkable document with an immense assigned task—the prevention of collisions between a vast array of vessels in a vast array of circumstances: vessels barely visible at 100 yards to vessels the size of horizontal skyscrapers; drifting without power or traveling at 30 knots or more; following unmarked lanes or crisscrossing open waters offering nothing more than an educated guess as to their intended course; in all conditions of weather, clear or fog, calm or storm; and often with no common language between their drivers.

But despite this enormous assignment, they do the job. Collisions can always be traced to at least one violation of the Rules. The key to avoiding further proof of this is a thorough understanding of the Rules and how to apply them, including the rules on what to do if an approaching vessel does not obey the rules.

This publication is not intended to teach the interpretation or application of the Rules, but merely to provide a copy of them. For valuable comments on the rules, the reader is directed to such texts as: *Farwell's Rules of the Nautical Road* by Richard A. Smith (7th edition, Naval Institute Press, 1994); *Handbook of the Nautical Rules of the Road* by C.B. Llana and G.P. Wisneskey (Naval Institute Press, 1991); *Nautical Rules of the Road* by B. A. Farnsworth and Larry C. Young (Cornell Maritime Press, 1990); *"RB-169" Navigation Rules* (Rules, comments, explanations, practice problems) by R. A. Block (Marine Education Textbooks, 1992); *Collisions and their Causes* by Richard A. Cahill (Fairplay Publications, 1983); *Collision Cases Judgments and Diagrams* by F.J. Buzek and H.M.C. Holdert (Lloyd's of London Press, 1990); *The Collision Regulations* by Richard B. Sturt (Lloyd's of London Press, 1991).

Skippers who travel between Inland and International waters, must know and use both sets of rules. Hopefully they will find the format of this presentation useful in that it emphasizes the differences between the two sets of rules.

Our only other contribution that might help with studying the Rules is the "Annotated Contents" provided. There is no index in the government edition of the Rules, so this added annotation is intended to help guide readers to a desired rule. The topic of each subsection of the rules has been noted for quick reference along with the page numbers for Inland and International Rules.

Aside from this expansion of the table of contents and a condensation of the Inland Rules, this book is a reformatted version (without illustrations) of the government edition of the rules (COMDTINST M16672.2B). We have updated this latest government edition for corrections presented in Notice to Mariners 16/91, 41/91, 33/92, 50/92, 27/93, and 41/94 — the notation means: week of the year/year. The International Rules are presented in full; the Inland Rules are presented only in those parts which differ from the International Rules. Areas where differences occur are clearly marked in the International Rules section of this book with margin bars. Note that the margin bars that appear in the government edition (COMDTINST M16672.2B) have a completely different meaning—in that edition margin bars are used to mark rules that have changed since the 1983 edition.

The International Rules can be read directly from the text presented here. **To obtain the full wording of the Inland Rules**, read the International Rules as if they applied to Inland Waters, but modify those individual sections that are marked by margin bars by referring to the Inland Rules section for the applicable inland rule.

In some cases, slight differences in wording between the International and Inland Rules represents important differences in meaning or application. These cases have been marked and included. In other cases, slight or even larger differences in wording do not appear to signify intended differences in ruling. In these cases, we have taken the liberty to treat these as identical rules and have not marked them.

David Burch
Starpath School of Navigation
Seattle, WA

Annotated Contents

PART A — GENERAL

Rule 1. Application

Rule 2. Responsibility

Rule 3. General Definitions

PART B — STEERING AND SAILING RULES

Section I — Conduct of Vessels in Any Condition of Visibility

Rule 4. Application

Rule 5. Look-Out

vi

Section II — Conduct of Vessels in Sight of One Another
See Rule 3 (k) and (l) for important definitions

Rule 31. Seaplanes

PART D — SOUND AND LIGHT SIGNALS

Rule 32. Definitions

Rule 33. Equipment for Sound Signals

Rule 34. Maneuvering and Warning Signals

Rule 35. Sound Signals in Restricted Visibility

Rule 36. Signals to Attract Attention

Rule 37. Distress Signals

PART E — Exemptions

Rule 38. Exemptions

INTRODUCTION
TO THE
INTERNATIONAL RULES

The International Rules in this book were formalized in the Convention on the International Regulations for Preventing Collisions at Sea, 1972, and became effective on July 15, 1977. The Rules (commonly called COLREGS) are part of the Convention, and vessels flying the flags of states ratifying the treaty are bound to the Rules. The United States has ratified this treaty and all United States flag vessels must adhere to these Rules where applicable. President Gerald R. Ford proclaimed 72 COLREGS and the Congress adopted them as the International Navigational Rules Act of 1977.

The 72 COLREGS were developed by the Inter-Governmental Maritime Consultative Organization (IMCO) which in May 1982 was renamed the International Maritime Organization (IMO). In November 1981, IMO's Assembly adopted 55 amendments to the 72 COLREGS which became effective on June 1, 1983. The IMO also adopted 9 more amendments which became effective on November 19, 1989. The International Rules in this book contain these amendments.

These Rules are applicable on waters outside of established navigational lines of demarcation. The lines are called COLREGS Demarcation Lines and delineate those waters upon which mariners shall comply with the Inland and International Rules. COLREGS Demarcation lines are contained in this book.

PART A—GENERAL

RULE 1. Application

(a) These Rules shall apply to all vessels upon the high seas and in all waters connected therewith navigable by seagoing vessels.

(b) Nothing in these Rules shall interfere with the operation of special rules made by an appropriate authority for roadsteads, harbors, rivers, lakes or inland waterways connected with the high seas and navigable by seagoing vessels. Such special rules shall conform as closely as possible to these Rules.

(c) Nothing in these Rules shall interfere with the operation of any special rules made by the Government of any State with respect to additional station or signal lights, shapes or whistle signals for ships of war and vessels proceeding under convoy, or with respect to additional station or signal lights or shapes for fishing vessels engaged in fishing as a fleet. These additional station or signal lights, shapes or whistle signals shall, so far as possible, be such that they cannot be mistaken for any light, shape or signal authorized elsewhere under these Rules[1].

(d) Traffic separation schemes may be adopted by the Organization for the purpose of these Rules.

(e) Whenever the Government concerned shall have determined that a vessel of special construction or purpose cannot comply fully with the provisions of any of these Rules with respect to the number, position, range or arc of visibility of lights or shapes, as well as to the disposition and characteristics of sound-signalling appliances, such vessel shall comply with such other provisions in regard to the number, position, range or arc of visibility of lights or shapes, as well as to the disposition and characteristic of sound-signalling appliances, as her Government shall have determined to be the closest possible compliance with these Rules in respect to that vessel.

See Inland part (f) on certificates of alternative compliance

---

[1] Submarines may display, as a distinctive means of identification, an intermittent flashing amber (yellow) beacon with a sequence of operation of one flash per second for three (3) seconds followed by a three (3) second off-period. Other special rules made by the Secretary of the Navy with respect to additional station and signal lights are found in Part 707 of Title 32, Code of Federal Regulations (32 CFR 707).

RULE 2. Responsibility

(a) Nothing in these Rules shall exonerate any vessel, or the owner, master or crew thereof, from the consequences of any neglect to comply with these Rules or of the neglect of any precaution which may be required by the ordinary practice of seamen, or by the special circumstances of the case.

(b) In construing and complying with these Rules due regard shall be had to all dangers of navigation and collision and to any special circumstances, including the limitations of the vessels involved, which may make a departure from these Rules necessary to avoid immediate danger.

RULE 3. General Definitions

For the purpose of these Rules, except where the context otherwise requires:

(a) The word "vessel" includes every description of water craft, including nondisplacement craft and seaplanes, used or capable of being used as a means of transportation on water.

(b) The term "power-driven vessel" means any vessel propelled by machinery.

(c) The term "sailing vessel" means any vessel under sail provided that propelling machinery, if fitted, is not being used.

(d) The term "vessel engaged in fishing" means any vessel fishing with nets, lines, trawls or other fishing apparatus which restrict maneuverability, but does not include a vessel fishing with trolling lines or other fishing apparatus which do not restrict maneuverability.

(e) The word "seaplane" includes any aircraft designed to maneuver on the water.

(f) The term "vessel not under command" means a vessel which through some exceptional circumstance is unable to maneuver as required by these Rules and is therefore unable to keep out of the way of another vessel.

(g) The term "vessel restricted in her ability to maneuver" means a vessel which from the nature of her work is restricted in her ability to maneuver as required by these Rules and is therefore unable to keep out of the way of another vessel.

The term 'vessels restricted in their ability to maneuver' shall include but not be limited to:

(i) a vessel engaged in laying, servicing or picking up a navigation mark, submarine cable or pipeline;

(ii) a vessel engaged in dredging, surveying or underwater operations;

(iii) a vessel engaged in replenishment or transferring persons, provisions or cargo while underway;

(iv) a vessel engaged in launching or recovery of aircraft;

(v) a vessel engaged in mine clearance operations;

(vi) a vessel engaged in a towing operation such as severely restricts the towing vessel and her tow in their ability to deviate from their course;

(h) The term "vessel constrained by her draft" means a power-driven vessel which, because of her draft in relation to the available depth and width of navigable water is severely restricted in her ability to deviate from the course she is following.

(i) The word "underway" means a vessel that is not at anchor, or made fast to shore, or aground.

(j) The words "length" and "breadth" of a vessel means her length overall and greatest breadth.

(k) Vessels shall be deemed to be in sight of one another only when one can be observed visually from the other.

(l) The term "restricted visibility" means any condition in which visibility is restricted by fog, mist, falling snow, heavy rainstorms, sandstorms or any other similar causes.

See Inland Rules for additional definitions on Inland waters.

PART B — STEERING AND SAILING RULES

Section I — Conduct of Vessels in Any Condition of Visibility

RULE 4. Application

Rules in this Section apply to any condition of visibility.

RULE 5. Look-out

Every vessel shall at all times maintain a proper look-out by sight and hearing as well as by all available means appropriate in the prevailing circumstances and conditions so as to make a full appraisal of the situation and of the risk of collision.

RULE 6. Safe Speed

Every vessel shall at all times proceed at a safe speed so that she can take proper and effective action to avoid collision and be stopped within a distance appropriate to the prevailing circumstances and conditions.

In determining a safe speed the following factors shall be among those taken into account:

(a) By all vessels:
    (i) the state of visibility;
    (ii) the traffic density including concentrations of fishing vessels or any other vessels;
    (iii) the maneuverability of the vessel with special reference to stopping distance and turning ability in the prevailing conditions;
    (iv) at night the presence of background light such as from shore lights or from back scatter of her own lights;
    (v) the state of wind, sea and current, and the proximity of navigational hazards;
    (vi) the draft in relation to the available depth of water.

(b) Additionally, by vessels with operational radar.
    (i) the characteristics, efficiency and limitations of the radar equipment;
    (ii) any constraints imposed by the radar range scale in use;
    (iii) the effect on radar detection of the sea state, weather and other sources of interference;
    (iv) the possibility that small vessels, ice and other floating objects may not be detected by radar at an adequate range;
    (v) the number, location and movement of vessels detected by radar;
    (vi) the more exact assessment of the visibility that may be possible when radar is used to determine the range of vessels or other objects in the vicinity.

RULE 7. Risk of Collision

(a) Every vessel shall use all available means appropriate to the prevailing circumstances and conditions to determine if risk of collision exists. If there is any doubt such risk shall be deemed to exist.

(b) Proper use shall be made of radar equipment if fitted and operational, including long-range scanning to obtain early warning of risk of collision and radar plotting or equivalent systematic observation of detected objects.

(c) Assumptions shall not be made on the basis of scanty information, especially scanty radar information.

(d) In determining if risk of collision exists the following considerations shall be among those taken into account:
  (i) such risk shall be deemed to exist if the compass bearing of an approaching vessel does not appreciably change.
  (ii) such risk may sometimes exist even when an appreciable bearing change is evident, particularly when approaching a very large vessel or a tow or when approaching a vessel at close range.

RULE 8. Action to Avoid Collision

(a) Any action taken to avoid collision shall, if the circumstances of the case admit, be positive, made in ample time and with due regard to the observance of good seamanship.

(b) Any alteration of course and/or speed to avoid collision shall, if the circumstances of the case admit, be large enough to be readily apparent to another vessel observing visually or by radar; a succession of small alterations of course and/or speed should be avoided.

(c) If there is sufficient sea room, alteration of course alone may be the most effective action to avoid a close-quarters situation provided that it is made in good time, is substantial and does not result in another close-quarters situation.

(d) Action taken to avoid collision with another vessel shall be such as to result in passing at a safe distance. The effectiveness of the action shall be carefully checked until the other vessel is finally past and clear.

(e) If necessary to avoid collision or allow more time to assess the situation, a vessel shall slacken her speed or take all way off by stopping or reversing her means of propulsion.

(f)   (i) A vessel which, by any of these rules, is required not to impede the passage or safe passage of another vessel shall, when required by the circumstances of the case, take early action to allow sufficient sea room for the safe passage of the other vessel.

(ii) A vessel required not to impede the passage or safe passage of another vessel is not relieved of this obligation if approaching the other vessel so as to involve risk of collision and shall, when taking action, have full regard to the action which may be required by the rules of this part.

(iii) A vessel, the passage of which is not to be impeded remains fully obliged to comply with the rules of this part when the two vessels are approaching one another so as to involve risk of collision.

RULE 9   Narrow Channels

(a) A vessel proceeding along the course of a narrow channel or fairway shall keep as near to the outer limit of the channel or fairway which lies on her starboard side as is safe and practicable.

(b) A vessel of less than 20 meters in length or a sailing vessel shall not impede the passage of a vessel which can safely navigate only within a narrow channel or fairway.

(c) A vessel engaged in fishing shall not impede the passage of any other vessel navigating within a narrow channel or fairway.

(d) A vessel shall not cross a narrow channel or fairway if such crossing impedes the passage of a vessel which can safely navigate only within such channel or fairway. The latter vessel may use the sound signal prescribed in Rule 34(d) if in doubt as to the intention of the crossing vessel.

(e)   (i) In a narrow channel or fairway when overtaking can take place only if the vessel to be overtaken has to take action to permit safe passing, the vessel intending to overtake shall indicate her intention by sounding the appropriate signal prescribed in Rule 34(c)(i). The vessel to be overtaken shall, if in agreement, sound the appropriate signal prescribed in Rule 34(c)(ii) and take steps to permit safe passing. If in doubt she may sound the signals prescribed in Rule 34(d).

(ii) This Rule does not relieve the overtaking vessel of her obligation under Rule 13.

(f) A vessel nearing a bend or an area of a narrow channel or fairway where other vessels may be obscured by an intervening obstruction shall navigate with particular alertness and caution and shall sound the appropriate signal prescribed in Rule 34(e).

—————— Means corresponding Inland Rule differs, see pages 64-73

(g) Any vessel shall, if the circumstances of the case admit, avoid anchoring in a narrow channel.

RULE 10 Traffic Separation Schemes

(a) This rule applies to traffic separation schemes adopted by the Organization and does not relieve any vessel of her obligation under any other rule.

(b) A vessel using a traffic separation scheme shall:
   (i) proceed in the appropriate traffic lane in the general direction of traffic flow for that lane;
   (ii) so far as practicable keep clear of a traffic separation line or separation zone;
   (iii) normally join or leave a traffic lane at the termination of the lane, but when joining or leaving from either side shall do so at as small an angle to the general direction of traffic flow as practicable.

(c) A vessel shall, so far as practicable, avoid crossing traffic lanes but if obliged to do so shall cross on a heading as nearly as practicable at right angles to the general direction of traffic flow.

(d) (i) A vessel shall not use an inshore traffic zone when she can safely use the apppropriate traffic lane within the adjacent traffic separation scheme. However, vessels of less than 20 meters in length, sailing vessels and vessels engaged in fishing may use the inshore traffic zone.
   (ii) Notwithstanding subparagraph (d)(i), a vessel may use an inshore traffic zone when en route to or from a port, offshore installation or structure, pilot station or any other place situated within the inshore traffic zone, or to avoid immediate danger.

(e) A vessel other than a crossing vessel or a vessel joining or leaving a lane shall not normally enter a separation zone or cross a separation line except:
   (i) in cases of emergency to avoid immediate danger;
   (ii) to engage in fishing within a separation zone.

(f) A vessel navigating in areas near the terminations of traffic separation schemes shall do so with particular caution.

(g) A vessel shall so far as practicable avoid anchoring in a traffic separation scheme or in areas near its terminations.

(h) A vessel not using a traffic separation scheme shall avoid it by as wide a margin as is practicable.

(i) A vessel engaged in fishing shall not impede the passage of any vessel following a traffic lane.

(j) A vessel of less than 20 meters in length or a sailing vessel shall not impede the safe passage of a power-driven vessel following a traffic lane.

(k) A vessel restricted in her ability to maneuver when engaged in an operation for the maintenance of safety of navigation in a traffic separation scheme is exempted from complying with this Rule to the extent necessary to carry out the operation.

(l) A vessel restricted in her ability to maneuver when engaged in an operation for the laying, servicing or picking up of a submarine cable, within a traffic separation scheme, is exempted from complying with this Rule to the extent necessary to carry out the operation.

Section II —Conduct of Vessels in Sight of One Another

RULE 11. Application

Rules in this Section apply to vessels in sight of one another.

RULE 12. Sailing Vessels

(a) when two sailing vessels are approaching one another, so as to involve risk of collision, one of them shall keep out of the way of the other as follows:
    (i) when each has the wind on a different side, the vessel which has the wind on the port side shall keep out of the way of the other;
    (ii) when both have the wind on the same side, the vessel which is to windward shall keep out of the way of the vessel which is to leeward;
    (iii) if a vessel with the wind on the port side sees a vessel to windward and cannot determine with certainty whether the other vessel has the wind on the port or on the starboard side, she shall keep out of the way of the other.

(b) For the purposes of this Rule the windward side shall be deemed to be the side opposite to that on which the mainsail is carried or, in the case of a square-rigged vessel, the side opposite to that on which the largest fore-and-aft sail is carried.

RULE 13. Overtaking

(a) Notwithstanding anything contained in the Rules of Part B, Sections I and II any vessel overtaking any other shall keep out of the way of the vessel being overtaken.

(b) A vessel shall be deemed to be overtaking when coming up with another vessel from a direction more than 22.5 degrees abaft her beam, that is, in such a position with reference to the vessel she is overtaking, that at night she would be able to see only the stern light of that vessel but neither of her sidelights.

(c) When a vessel is in any doubt as to whether she is overtaking another, she shall assume that this is the case and act accordingly.

(d) Any subsequent alteration of the bearing between the two vessels shall not make the overtaking vessel a crossing vessel within the meaning of these Rules or relieve her of the duty of keeping clear of the overtaken vessel until she is finally past and clear.

RULE 14. Head-on Situation

(a) When two power-driven vessels are meeting on reciprocal or nearly reciprocal courses so as to involve risk of collision each shall alter her course to starboard so that each shall pass on the port side of the other.

(b) Such a situation shall be deemed to exist when a vessel sees the other ahead or nearly ahead and by night she could see the masthead lights of the other in a line or nearly in a line and/or both side lights and by day she observes the corresponding aspect of the other vessel.

(c) When a vessel is in any doubt as to whether such a situation exists she shall assume that it does exist and act accordingly.

See Inland Rules on vessels crossing rivers.

RULE 15. Crossing Situations

When two power-driven vessels are crossing so as to involve risk of collision, the vessel which has the other on her own starboard side shall keep out of the way and shall, if the circumstances of the case admit, avoid crossing ahead of the other vessel.

See Inland Rules on vessels crossing rivers.

RULE 16. Action by Give-way Vessel

Every vessel which is directed to keep out of the way of another vessel shall, so far as possible, take early and substantial action to keep well clear.

RULE 17. Action by Stand-on Vessel

(a) (i) Where one of two vessels is to keep out of the way, the other shall keep her course and speed.

(ii) The latter vessel may, however, take action to avoid collision by her maneuver alone, as soon as it becomes apparent to her that the vessel required to keep out of the way is not taking appropriate action in compliance with these Rules.

(b) When, from any cause, the vessel required to keep her course and speed finds herself so close that collision cannot be avoided by the action of the give-way vessel alone, she shall take such action as will best aid to avoid collision.

(c) A power-driven vessel which takes action in a crossing situation in accordance with subparagraph (a)(ii) of this Rule to avoid collision with another power-driven vessel shall, if the circumstances of the case admit, not alter course to port for a vessel on her own port side.

(d) This Rule does not relieve the give-way vessel of her obligation to keep out of the way.

RULE 18. Responsibilities Between Vessels

Except where Rules 9, 10, and 13 otherwise require:

(a) A power-driven vessel underway shall keep out of the way of:
   (i) a vessel not under command;
   (ii) a vessel restricted in her ability to maneuver;
   (iii) a vessel engaged in fishing;
   (iv) a sailing vessel.

(b) A sailing vessel underway shall keep out of the way of:
   (i) a vessel not under command;
   (ii) a vessel restricted in her ability to maneuver;
   (iii) a vessel engaged in fishing.

(c) A vessel engaged in fishing when underway shall, so far as possible, keep out of the way of:
   (i) a vessel not under command;
   (ii) a vessel restricted in her ability to maneuver.

(d) (i) Any vessel other than a vessel not under command or a vessel restricted in her ability to maneuver shall, if the circumstances of the case admit, avoid impeding the safe passage of a vessel constrained by her draft, exhibiting the signals in Rule 28.

(ii) A vessel constrained by her draft shall navigate with particular caution having full regard to her special condition.

(e) A seaplane on the water shall, in general, keep well clear of all vessels and avoid impeding their navigation. In circumstances, however, where risk of collision exists, she shall comply with the Rules of this Part.

Section III — Conduct of Vessels in Restricted Visibility

RULE 19. Conduct of Vessels in Restricted Visibility

(a) This Rule applies to vessels not in sight of one another when navigating in or near an area of restricted visibility.

(b) Every vessel shall proceed at a safe speed adapted to the prevailing circumstances and conditions of restricted visibility. A power-driven vessel shall have her engines ready for immediate maneuver.

(c) Every vessel shall have due regard to the prevailing circumstances and conditions of restricted visibility when complying with the Rules of Section I of this Part.

(d) A vessel which detects by radar alone the presence of another vessel shall determine if a close-quarters situation is developing and/or risk of collision exists. If so, she shall take avoiding action in ample time, provided that when such action consists of an alteration of course, so far as possible the following shall be avoided:
(i) an alteration of course to port for a vessel forward of the beam, other than for a vessel being overtaken;
(ii) an alteration of course toward a vessel abeam or abaft the beam.

(e) Except where it has been determined that a risk of collision does not exist, every vessel which hears apparently forward of her beam the fog signal of another vessel, or which cannot avoid a close-quarters situation with another vessel forward of her beam, shall reduce her speed to the minimum at which she can be kept on her course. She shall if necessary take all her way off and in any event navigate with extreme caution until danger of collision is over.

Means corresponding Inland Rule differs, see pages 64-73

PART C — LIGHTS AND SHAPES

RULE 20. Application

(a) Rules in this Part shall be complied with in all weathers.

(b) The Rules concerning lights shall be complied with from sunset to sunrise, and during such times no other lights shall be exhibited, except such lights as cannot be mistaken for the lights specified in these Rules or do not impair their visibility or distinctive character, or interfere with the keeping of the proper look-out.

(c) The lights prescribed by these Rules shall, if carried, also be exhibited from sunrise to sunset in restricted visibility and may be exhibited in all other circumstances when it is deemed necessary.

(d) The Rules concerning shapes shall be complied with by day.

(e) The lights and shapes specified in these Rules shall comply with the provisions of Annex I to these Regulations.

RULE 21. Definitions

(a) "Masthead light" means a white light placed over the fore and aft centerline of the vessel showing an unbroken light over an arc of the horizon of 225 degrees and so fixed as to show the light from right ahead to 22.5 degrees abaft the beam on either side of the vessel.

(b) "Sidelights" means a green light on the starboard side and a red light on the port side each showing an unbroken light over an arc of the horizon of 112.5 degrees and so fixed as to show the light from right ahead to 22.5 degrees abaft the beam on its respective side. In a vessel of less than 20 meters in length the sidelights may be combined in one lantern carried on the fore and aft centerline of the vessel.

(c) "Sternlight" means a white light placed as nearly as practicable at the stern showing an unbroken light over an arc of the horizon of 135 degrees and so fixed as to show the light 67.5 degrees from right aft on each side of the vessel.

(d) "Towing light" means a yellow light having the same characteristics as the "sternlight" defined in paragraph (c) of this Rule.

(e) "All-round light" means a light showing an unbroken light over an arc of the horizon of 360 degrees.

(f) "Flashing light" means a light flashing at regular intervals at a frequency of 120 flashes or more per minute.

—————— Means corresponding Inland Rule differs, see pages 64-73

(g) *See Inland Rules for use of special flashing light.*

RULE 22. Visibility of Lights

The lights prescribed in these Rules shall have an intensity as specified in Section 8 of Annex I to these Regulations so as to be visible at the following minimum ranges:

(a) In vessels of 50 meters or more in length:
    —a masthead light, 6 miles;
    —a sidelight, 3 miles;
    —a sternlight, 3 miles;
    —a towing light, 3 miles;
    —a white, red, green, or yellow all-round light, 3 miles.

(b) In vessels of 12 meters or more in length but less than 50 meters in length:
    —a masthead light, 5 miles;
    except that where the length of the vessel is less than 20 meters, 3 miles;
    —a sidelight, 2 miles;
    —a sternlight, 2 miles;
    —a towing light, 2 miles;
    —a white, red, green or yellow all-round light, 2 miles.

(c) In vessels of less than 12 meters in length:
    —a masthead light, 2 miles;
    —a sidelight, 1 mile;
    —a sternlight, 2 miles;
    —a towing light, 2 miles;
    —a white, red, green or yellow all-round light, 2 miles.

(d) In inconspicuous, partly submerged vessels or objects being towed:
    —a white all-round light, 3 miles.

RULE 23. Power-driven Vessels Underway

(a) A power-driven vessel underway shall exhibit:
    (i) a masthead light forward;
    (ii) a second masthead light abaft of and higher than the forward one; except that a vessel of less than 50 meters in length shall not be obliged to exhibit such light but may do so;
    (iii) sidelights;   (iv) a sternlight.

(b) an air-cushion vessel when operating in the non-displacement mode shall, in addition to the lights prescribed in paragraph (a) of this Rule, exhibit an all-round flashing yellow light.

(c) (i) A power-driven vessel of less than 12 meters in length may in lieu of the lights prescribed in paragraph (a) of this Rule exhibit an all-round white light and sidelights;

(ii) a power-driven vessel of less than 7 meters in length whose maximum speed does not exceed 7 knots may in lieu of the lights prescribed in paragraph (a) of his Rule exhibit an all-round white light and shall, if practicable, also exhibit sidelights;

(iii) the masthead light or all-round white light on a power-driven vessel of less than 12 meters in length may be displaced from the fore and aft centerline of the vessel if centerline fitting is not practicable, provided that the sidelights are combined in one lantern which shall be carried on the fore and aft centerline of the vessel or located as nearly as practicable in the same fore and aft line as the masthead light or the all-round white light.

(d) *See Inland Rules for Great Lakes*

RULE 24. Towing and Pushing

(a) A power-driven vessel when towing shall exhibit:

(i) instead of the light prescribed in Rule 23 (a)(i) or (a)(ii), two masthead lights in a vertical line. When the length of the tow, measuring from the stern of the towing vessel to the after end of the tow exceeds 200 meters, three such lights in a vertical line;

(ii) sidelights;

(iii) a sternlight;

(iv) a towing light in a vertical line above the sternlight;

(v) when the length of the tow exceeds 200 meters, a diamond shape where it can best be seen.

(b) When a pushing vessel and a vessel being pushed ahead are rigidly connected in a composite unit they shall be regarded as a power-driven vessel and exhibit the lights prescribed in Rule 23.

(c) A power-driven vessel when pushing ahead or towing alongside, except in the case of a composite unit, shall exhibit;

(i) instead of the light prescribed in Rule 23(a)(i) or (a)(ii), two masthead lights in a vertical line;

(ii) sidelights;

(iii) a sternlight.

(d) A power-driven vessel to which paragraphs (a) or (c) of this Rule apply shall also comply with Rule 23(a)(ii).

(e) A vessel or object being towed, other than those mentioned in paragraph (g) of this Rule shall exhibit;

Means corresponding Inland Rule differs, see pages 64-73

(i) sidelights;

(ii) a sternlight;

(iii) when the length of the tow exceeds 200 meters, a diamond shape where it can best be seen.

(f) Provided that any number of vessels being towed alongside or pushed in a group shall be lighted as one vessel:

(i) a vessel being pushed ahead, not being part of a composite unit, shall exhibit at the forward end, sidelights;

(ii) a vessel being towed alongside shall exhibit a sternlight and at the forward end, sidelights.

(g) An inconspicuous, partly submerged vessel or object, or combination of such vessels or objects being towed, shall exhibit:

(i) if it is less than 25 meters in breadth, one all-round white light at or near the forward end and one at or near the after end except that dracones need not exhibit a light at or near the forward end;

(ii) if it is 25 meters of more in breadth, two additional all-round white lights at or near the extremities of its breadth;

(iii) if it exceeds 100 meters in length, additional all-round white lights between the lights prescribed in subparagraphs (i) and (ii) so that the distance between the lights shall not exceed 100 meters;

(iv) a diamond shape at or near the aftermost extremity of the last vessel or object being towed and if the length of the tow exceeds 200 meters an additional diamond shape where it can best be seen and located as far forward as is practicable.

(h) Where from any sufficient cause it is impracticable for a vessel or object being towed to exhibit the lights or shapes prescribed in paragraph (e) or (g) of this Rule, all possible measures shall be taken to light the vessel or object towed or at least to indicate the presence of such vessel or object.

(i) Where from any sufficient cause it is impracticable for a vessel not normally engaged in towing operations to display the lights prescribed in paragraph (a) or (c) of this Rule, such vessel shall not be required to exhibit those lights when engaged in towing another vessel in distress or otherwise in need of assistance. All possible measures shall be taken to indicate the nature of the relationship between the towing vessel and the vessel being towed as authorized by Rule 36, in particular by illuminating the towline.

RULE 25. Sailing Vessels Underway and Vessels Under Oars

(a) A sailing vessel underway shall exhibit:

(i) sidelights;   (ii) a sternlight.

(b) In a sailing vessel of less than 20 meters in length the lights prescribed in paragraph (a) of this Rule may be combined in one lantern carried at or near the top of the mast where it can best be seen.

(c) A sailing vessel underway may, in addition to the lights prescribed in paragraph (a) of this Rule, exhibit at or near the top of the mast, where they can best be seen, two all-round lights in a vertical line, the upper being red and the lower green, but these lights shall not be exhibited in conjunction with the combined lantern permitted by paragraph (b) of this Rule.

(d) (i) A sailing vessel of less than 7 meters in length shall, if practicable, exhibit the lights prescribed in paragraph (a) or (b) of this Rule, but if she does not, she shall have ready at hand an electric torch or lighted lantern showing a white light which shall be exhibited in sufficient time to prevent collision.
(ii) A vessel under oars may exhibit the lights prescribed in this Rule for sailing vessels, but if she does not, she shall have ready at hand an electric torch or lighted lantern showing a white light which shall be exhibited in sufficient time to prevent collision.

(e) A vessel proceeding under sail when also being propelled by machinery shall exhibit forward where it can best be seen a conical shape, apex downwards.

RULE 26. Fishing Vessels

(a) A vessel engaged in fishing, whether underway or at anchor, shall exhibit only the lights and shapes prescribed in this Rule.

(b) A vessel when engaged in trawling, by which is meant the dragging through the water of a dredge net or other apparatus used as a fishing appliance, shall exhibit:
(i) two all-round lights in a vertical line, the upper being green and the lower white, or a shape consisting of two cones with their apexes together in a vertical line one above the other; a vessel of less than 20 meters in length may instead of this shape exhibit a basket;
(ii) a masthead light abaft of and higher than the all-round green light; a vessel of less than 50 meters in length shall not be obliged to exhibit such a light but may do so;
(iii) when making way through the water, in addition to the lights prescribed in this paragraph, sidelights and a sternlight.

(c) A vessel engaged in fishing, other than trawling, shall exhibit:
(i) two all-round lights in a vertical line, the upper being red and the lower white, or a shape consisting of two cones with apexes together in a vertical line one above the other; a vessel of less than 20 meters in length may instead of

this shape exhibit a basket;

(ii) when their is outlying gear extending more than 150 meters horizontally from the vessel, an all-round white light or a cone apex upwards in the direction of the gear;

(iii) when making way through the water, in addition to the lights prescribed in this paragraph, sidelights and a sternlight.

(d) A vessel engaged in fishing in close proximity to other vessels engaged in fishing may exhibit the additional signals described in Annex II to these Regulations.

(e) A vessel when not engaged in fishing shall not exhibit the lights or shapes prescribed in this Rule, but only those prescribed for a vessel of her length.

RULE 27. Vessels Not Under Command or Restricted in their Ability to Maneuver

(a) A vessel not under command shall exhibit:

(i) two all-round red lights in a vertical line where they can best be seen;

(ii) two balls or similar shapes in a vertical line where they can best be seen;

(iii) when making way through the water, in addition to the lights prescribed in this paragraph, sidelights and a sternlight.

(b) A vessel restricted in her ability to maneuver, except a vessel engaged in mineclearance operations, shall exhibit:

(i) three all-round lights in a vertical line where they can best be seen. The highest and lowest of these lights shall be red and the middle light shall be white;

(ii) three shapes in a vertical line where they can best be seen. The highest and lowest of these shapes shall be balls and the middle one a diamond;

(iii) when making way through the water, a masthead light or lights, sidelights, and a sternlight, in addition to the lights prescribed in subparagraph (i);

(iv) when at anchor, in addition to the lights or shapes prescribed in subparagraphs (i) and (ii), the light, lights or shape prescribed in Rule 30.

(c) A power-driven vessel engaged in a towing operation such as severely restricts the towing vessel and her tow in their ability to deviate from their course shall, in addition to the lights or shapes prescribed in Rule 24(a), exhibit the lights or shapes prescribed in subparagraphs (b)(i) and (ii) of this Rule.

(d) A vessel engaged in dredging or underwater operations, when restricted in her ability to maneuver, shall exhibit the lights and shapes prescribed in subparagraphs (b)(i), (ii), and (iii) of this Rule and shall in addition, when an obstruction exits, exhibit:

(i) two all-round red lights or two balls in a vertical line to indicate the side on which the obstruction exists;

(ii) two all-round green lights or two diamonds in a vertical line to indicate the side on which another vessel may pass;

(iii) when at anchor, the lights or shapes prescribed in this paragraph instead of the lights or shape prescribed in Rule 30.

(e) Whenever the size of a vessel engaged in diving operations makes it impracticable to exhibit all lights and shapes prescribed in paragraph (d) of this Rule, the following shall be exhibited:

(i) three all-round lights in a vertical line where they can best be seen. The highest and lowest of these lights shall be red and the middle light shall be white.

(ii) a rigid replica of the International Code flag "A" not less than 1 meter in height. Measures shall be taken to ensure its all-round visibility.

(f) A vessel engaged in mineclearance operations shall in addition to the lights prescribed for a power-driven vessel in Rule 23 or to the lights or shape prescribed for a vessel at anchor in Rule 30 as appropriate, exhibit three all-round green lights or three balls. One of these lights or shapes shall be exhibited near the foremast head and one at each end of the fore yard. These lights or shapes indicate that it is dangerous for another vessel to approach within 1000 meters of the mineclearance vessel.

(g) Vessels of less than 12 meters in length, except those engaged in diving operations, shall not be required to exhibit the lights and shapes prescribed in this Rule.

(h) The signals prescribed in this Rule are not signals of vessels in distress and requiring assistance. Such signals are contained in Annex IV to these Regulations.

RULE 28. Vessels Constrained by their Draft

A vessel constrained by her draft may, in addition to the lights prescribed for power-driven vessels in Rule 23, exhibit where they can best be seen three all-round red lights in a vertical line, or a cylinder.

RULE 29. Pilot vessels

(a) A vessel engaged on pilotage duty shall exhibit:

(i) at or near the masthead, two all-round lights in a vertical line, the upper being white and the lower red;

(ii) when underway, in addition, sidelights and a sternlight;

(iii) when at anchor, in addition to the lights prescribed in subparagraph (i), the light, lights or shape prescribed in Rule 30 for vessels at anchor.

(b) A pilot vessel when not engaged on pilotage duty shall exhibit the lights or shapes prescribed for a similar vessel of her length.

RULE 30. Anchored Vessels and Vessels Aground

(a) A vessel at anchor shall exhibit where it can best be seen:
    (i) in the fore part, an all-round white light or one ball;
    (ii) at or near the stern and at a lower level than the light prescribed in subparagraph (i) an all-round white light.

(b) A vessel of less than 50 meters in length may exhibit an all-round white light where it can best be seen instead of the lights prescribed in paragraph (a) of this Rule.

(c) A vessel at anchor may, and a vessel of 100 meters and more in length shall, also use the available working or equivalent lights to illuminate her decks.

(d) A vessel aground shall exhibit the lights prescribed in paragraph (a) or (b) of this Rule and in addition, where they can best be seen:
    (i) two all-round red lights in a vertical line;
    (ii) three balls in a vertical line.

(e) A vessel of less than 7 meters in length, when at anchor, not in or near a narrow channel, fairway or anchorage, or where other vessels normally navigate, shall not be required to exhibit the lights or shapes prescribed in paragraphs (a) and (b) of this Rule.

(f) A vessel of less than 12 meters in length, when aground, shall not be required to exhibit the lights or shapes prescribed in subparagraphs (d)(i) and (ii) of this Rule.

(g) *See Inland Rules for exemptions in special anchorages.*

RULE 31. Seaplanes

Where it is impracticable for a seaplane to exhibit lights and shapes of the characteristics or in the positions prescribed in the Rules of this Part she shall exhibit lights and shapes as closely similar in characteristics and position as is possible.

PART D — SOUND AND LIGHT SIGNALS

RULE 32. Definitions

(a) The word "whistle" means any sound signaling appliance capable of producing the prescribed blasts and which complies with the specifications in Annex III to these Regulations.

(b) The term "short blast" means a blast of about one second's duration.

(c) The term "prolonged blast" means a blastof from four to six seconds' duration.

RULE 33. Equipment for Sound Signals

(a) A vessel of 12 meters or more in length shall be provided with a whistle and a bell and a vessel of 100 meters or more in length shall, in addition, be provided with a gong, the tone and sound of which cannot be confused with that of the bell. The whistle, bell and gong shall comply with the specifications in Annex III to these Regulations. The bell or gong or both may be replaced by other equipment having the same respective sound characteristics, provided that manual sounding of the prescribed signals shall always be possible.

(b) A vessel of less than 12 meters in length shall not be obliged to carry the sound signaling appliances prescribed in paragraph (a) of this Rule but if she does not, she shall be provided with some other means of making an efficient sound signal.

RULE 34. Maneuvering and Warning Signals

(a) When vessels are in sight of one another, a power-driven vessel underway, when maneuvering as authorized or required by these Rules, shall indicate that maneuver by the following signals on her whistle.

— one short blast to mean "I am altering my course to starboard";

— two short blasts to mean "I am altering my course to port";

— three short blasts to mean "I am operating astern propulsion".

(b) Any vessel may supplement the whistle signals prescribed in paragraph (a) of this Rule by light signals, repeated as appropriate, whilst the maneuver is being carried out:

(i) these light signals shall have the following significance:

— one flash to mean "I am altering my course to starboard";

▬▬▬▬▬ Means corresponding Inland Rule differs, see pages 64-73

— two flashes to mean " I am altering my course to port";

— three flashes to mean "I am operating astern propulsion";

(ii) the duration of each flash shall be about one second, the interval between flashes shall be about one second, and the interval between successive signals shall be not less than 10 seconds;

(iii) the light used for this signal shall, if fitted, be an all-round white light, visible at a minimum range of 5 miles, and shall comply with the provisions of Annex I to these Regulations.

(c) When in sight of one another in a narrow channel or fairway:

(i) a vessel intending to overtake another shall in compliance with Rulc(9)(c)(i) indicate her intention by the following signals on her whistle:

— two prolonged blasts followed by one short blast to mean "I intend to overtake you on your starboard side";

— two prolonged blasts followed by two short blasts to mean "I intend to overtake you on our port side".

(ii) The vessel about to be overtaken when acting in accordance with Rule 9 (e)(i) shall indicate her agreement by the following signal on her whistle:

— one prolonged, one short, one prolonged and one short blast, in that order.

(d) When vessels in sight of one another are approaching each other and from any cause either vessel fails to understand the intentions or actions of the other, or is in doubt whether sufficient action is being taken by the other to avoid collision, the vessel in doubt shall immediately indicate such doubt by giving at least five short and rapid blasts on the whistle. Such signal may be supplemented by a light signal of at least five short and rapid flashes.

(e) A vessel nearing a bend or an area of a channel or fairway where other vessels may be obscured by an intervening obstruction shall sound one prolonged blast. Such signal shall be answered with a prolonged blast by any approaching vessel that may be within hearing around the bend or behind the intervening obstruction. (f) If whistles are fitted on a vessel at a distance apart of more than 100 meters, one whistle only shall be used for giving maneuvering and warning signals.

See Inland Rules for signals leaving a dock and on radiotelephone usage.

————— Means corresponding Inland Rule differs, see pages 64-73

RULE 35. Sound Signals in Restricted Visibility

In or near an area of restricted visibility, whether by day or night, the signals prescribed in this Rule shall be used as follows:

(a) A power-driven vessel making way through the water shall sound at intervals of not more than 2 minutes one prolonged blast.

(b) A power-driven vessel underway but stopped and making no way through the water shall sound at intervals of not more than 2 minutes two prolonged blasts in succession with an interval of about 2 seconds between them.

(c) A vessel not under command, a vessel restricted in her ability to maneuver, a vessel constrained by her draft, a sailing vessel, a vessel engaged in fishing and a vessel engaged in towing or pushing another vessel shall, instead of the signals prescribed in paragraphs (a) or (b) of this Rule, sound at intervals of not more than 2 minutes three blasts in succession, namely one prolonged followed by two short blasts.

(d) A vessel engaged in fishing, when at anchor, and a vessel restricted in her ability to maneuver when carrying out her work at anchor, shall instead of the signals prescribed in paragraph (g) of this Rule sound the signal prescribed in paragraph (c) of this Rule.

(e) A vessel towed or if more than one vessel is towed the last vessel of the tow, if manned, shall at intervals of not more than 2 minutes sound four blasts in succession, namely one prolonged followed by three short blasts. When practicable, this signal shall be made immediately after the signal made by the towing vessel.

(f) When a pushing vessel and a vessel being pushed ahead are rigidly connected in a composite unit they shall be regarded as a power-driven vessel and shall give the signals prescribed in paragraphs (a) or (b) of this Rule.

(g) A vessel at anchor shall at intervals of not more than one minute ring the bell rapidly for about 5 seconds. In a vessel of 100 meters or more in length the bell shall be sounded in the forepart of the vessel and immediately after the ringing of the bell the gong shall be sounded rapidly for about 5 seconds in the after part of the vessel. A vessel at anchor may in addition sound three blasts in succession, namely one short, one prolonged and one short blast, to give warning of her position and of the possibility of collision to an approaching vessel.

(h) A vessel aground shall give the bell signal and if required the gong signal prescribed in paragraph (g) of this Rule and shall, in addition, give three separate and distinct strokes on the bell immediately before and after the rapid ringing of the bell. A vessel aground may in addition sound an appropriate whistle signal.

───── Means corresponding Inland Rule differs, see pages 64-73

(i) A vessel less than 12 meters in length shall not be obliged to give the above-mentioned signals but, if she does not, shall make some other efficient sound signal at intervals of not more than 2 minutes.

(j) A pilot vessel when engaged on pilotage duty may in addition to the signals prescribed in paragraphs (a), (b), or (g) of this Rule sound an identity signal consisting of four short blasts.

See Inland Rule part (j) for exemptions in special anchorages.

RULE 36. Signals to Attract Attention

If necessary to attract attention of another vessel, any vessel may make light or sound signals that cannot be mistaken for any signal authorized elsewhere in these Rules, or may direct the beam of her searchlight in the direction of the danger, in such a way as not to embarrass any vessel. Any light to attract the attention of another vessel shall be such that it cannot be mistaken for any aid to navigation. For the purpose of this Rule the use of high intensity intermittent or revolving lights, such as strobe lights, shall be avoided.

RULE 37. Distress Signals

When a vessel is in distress and requires assistance she shall use or exhibit the signals described in Annex IV to these regulations.

PART E — EXEMPTIONS

RULE 38. Exemptions

Any vessel (or class of vessels) provided that she complies with the requirements of the International Regulations for Preventing Collisions at Sea, 1960, the keel of which is laid or which is at a corresponding stage of construction before the entry into force of these Regulations may be exempted from compliance therewith as follows:

(a) The installation of lights with ranges prescribed in Rule 22, until four years after the date of entry into force of these Regulations.

(b) The installation of lights with color specifications as prescribed in Section 7 of Annex I to these Regulations, until four years after the date of entry into force of these Regulations.

(c) The repositioning of lights as a result of conversion from Imperial to metric units and rounding off measurement figures, permanent exemption.

(d) (i) The repositioning of masthead lights on vessels of less than 150 meters in length, resulting from the prescriptions of Section 3(a) of Annex I to these Regulations, permanent exemption.

(ii) The repositioning of masthead lights on vessels of 150 meters or more in length, resulting from the prescriptions of Section 3(a) of Annex I to these Regulations, until 9 years after the date of entry into force of these Regulations.

(e) The repositioning of masthead lights resulting from the prescriptions of Section 2(b) of Annex I to these Regulations, until 9 years after the date of entry into force of these Regulations.

(f) The repositioning of sidelights resulting from the prescriptions of Sections 2(g) and 3(b) of Annex I to these Regulations, until 9 years after the date of entry into force of these Regulations.

(g) The requirements for sound signal appliances prescribed in Annex III to these Regulations, until 9 years after the date of entry into force of these Regulations.

(h) The repositioning of all-round lights resulting from the prescription of Section 9(b) of Annex I to these Regulations, permanent exemption.

ANNEX I
Positioning and Technical Details of Lights and Shapes

1. Definition
The term "height above the hull" means height above the uppermost continuous deck. This height shall be measured from the position vertically beneath the location of the light.

2. Vertical positioning and spacing of lights
(a) On a power-driven vessel of 20 meters or more in length the masthead lights shall be placed as follows:

(i) the forward masthead light, or if only one masthead light is carried, then that light, at a height above the hull of not less than 6 meters, and, if the breadth of the vessel exceeds 6 meters, then at a height above the hull not less than such breadth, so however that the light need not be placed at a greater height above the hull than 12 meters;

(ii) when two masthead lights are carried the after one shall be at least 4.5 meters vertically higher than the forward one.

(b) The vertical separation of masthead lights of power-driven vessels shall be such that in all normal conditions of trim the after light will be seen over and separate from the forward light at a distance of 1000 meters from the stem when viewed from sea level.

(c) The masthead light of a power-driven vessel of 12 meters but less than 20 meters in length shall be placed at a height above the gunwale of not less than 2.5 meters.

(d) A power-driven vessel of less than 12 meters in length may carry the uppermost light at a height of less than 2.5 meters above the gunwale. When however a masthead light is carried in addition to sidelights and a sternlight or the all-round light prescribed on Rule 23(c)(i) is carried in addition to sidelights, then such masthead light or all-round light shall be carried at least 1 meter higher than the sidelights.

(e) One of the two or three masthead lights prescribed for a power-driven vessel when engaged in towing or pushing another vessel shall be placed in the same position as either the forward masthead light or the after masthead light; provided that, if carried on the aftermast, the lowest after masthead light shall be at least 4.5 meters vertically higher than the forward masthead light.

(f) (i) The masthead light or lights prescribed in Rule 23(a) shall be so placed as to be above and clear of all other lights and obstructions except as described in subparagraph (ii).

(ii) When it is impracticable to carry the all-round lights prescribed by Rule 27(b)(i) or Rule 28 below the masthead lights, they may be carried above the after masthead light(s) or vertically in between the forward masthead light(s) and after masthead light(s) provided that in the latter case the requirements of Section 3(c) of this Annex shall be complied with.

(g) The sidelights of a power-driven vessel shall be placed at a height above the hull not greater than three quarters of that of the forward masthead light. They shall not be so low as to be interfered with by deck lights.

(h) The sidelights, if in a combined lantern and carried on a power-driven vessel of less than 20 meters in length, shall be placed not less than 1 meter below the masthead light.

(i) When the Rules prescribe two or three lights to be carried in a vertical line, they shall be spaced as follows:

(i) on a vessel of 20 meters in length or more such lights shall be spaced not less than 2 meters apart, and the lowest of these lights shall, except where a towing light is required, be placed at a height of not less than 4 meters above the hull;

(ii) on a vessel of less than 20 meters in length such lights shall be spaced not less than 1 meter apart and the lowest of these lights shall, except where a towing light is required, be placed at a height of not less than 2 meters above the gunwale;

(iii) when three lights are carried they shall be equally spaced.

(j) The lower of the two all-round lights prescribed for a vessel when engaged in fishing shall be at a height above the sidelights not less than twice the distance between the two vertical lights.

(k) The forward anchor light prescribed in Rule 30(a)(i), when two are carried, shall not be less than 4.5 meters above the after one. On a vessel of 50 meters or more in length this forward anchor light shall be placed at a height of not less than 6 meters above the hull.

3. Horizontal positioning and spacing of lights
(a) When two masthead lights are prescribed for a power-driven vessel, the horizontal distance between them shall not be less than one half of the length of the vessel but need not be more than 100 meters. The forward light shall be placed not more than one quarter of the length of the vessel from the stem.

Means corresponding Inland Annex differs, see pages 74-84

(b) On a power-driven vessel of 20 meters or more in length the sidelights shall not be placed in front of the forward masthead lights. They shall be placed at or near the side of the vessel.

(c) When the lights prescribed in Rule 27(b)(i) or Rule 28 are placed vertically between the forward masthead light(s) and the after masthead light(s) these all-round lights shall be placed at a horizontal distance of not less than 2 meters from the fore and aft centerline of the vessel in the athwartship direction.

4. Details of location of direction-indicating lights for fishing vessels, dredgers and vessels engaged in underwater operations

(a) The light indicating the direction of the outlying gear from a vessel engaged in fishing as prescribed in Rule 26(c)(ii) shall be placed at a horizontal distance of not less than 2 meters and not more than 6 meters away from the two all-round red and white lights. This light shall be placed not higher than the all-round white light prescribed in Rule 26(c)(i) and not lower than the sidelights.

(b) The lights and shapes on a vessel engaged in dredging or underwater operations to indicate the obstructed side and/or the side on which it is safe to pass, as prescribed in Rule 27(d)(i) and (ii), shall be placed at the maximum practical horizontal distance, but in no case less than 2 meters, from the lights or shapes prescribed in Rule 27(b)(i) and (ii). In no case shall the upper of the these lights or shapes be at a greater height than the lower of the three lights or shapes prescribed in Rule 27(b)(i) and (ii)

5. Screens for sidelights

The sidelights of vessels of 20 meters or more in length shall be fitted with inboard screens painted matt black, and meeting the requirements of Section 9 of this Annex. On vessels of less than 20 meters in length the sidelights, if necessary to meet the requirements of Section 9 of this Annex, shall be fitted with inboard matt black screens. With a combined lantern, using a single vertical filament and a very narrow division between the green and red sections, external screens need not be fitted.

6. Shapes

(a) Shapes shall be black and of the following sizes:

(i) a ball shall have a diameter of not less than 0.6 meter;

(ii) a cone shall have a base diameter of not less than 0.6 meter and a height equal to its diameter;

(iii) a cylinder shall have a diameter of at least 0.6 meter and a height of twice its diameter;

(iv) a diamond shape shall consist of two cones as defined in (ii) above having a common base.

(b) The vertical distance between shapes shall be at least 1.5 meters.

(c) In a vessel of less than 20 meters in length shapes of lesser dimensions but commensurate with the size of the vessel may be used and the distance apart may be correspondingly reduced.

7. Color specification of lights

The chromaticity of all navigation lights shall conform to the following standards, which lie within the boundaries of the area of the diagram specified for each color by the International Commission on Illumination (CIE). The boundaries of the area for each color are given by indicating the corner coordinates, which are as follows:

(i) White:
x 0.525 0.525 0.452 0.310 0.310 0.443
y 0.382 0.440 0.440 0.348 0.283 0.382

(ii) Green:
x 0.028 0.009 0.300 0.203
y 0.385 0.723 0.511 0.356

(iii) Red:
x 0.680 0.660 0.735 0.721
y 0.320 0.320 0.265 0.259

(iv) Yellow
x 0.612 0.618 0.575 0.575
y 0.382 0.382 0.425 0.406

8. Intensity of lights

(a) The minimum luminous intensity of lights shall be calculated by using the formula:

$$I = 3.43 \times 10^6 \times T \times D^2 \times K^{-D}$$

where:
I is luminous intensity in candelas under service conditions,
T is threshold factor 2×10^{-7} lux,
D is range of visibility (luminous range) of the light in nautical miles,
K is atmospheric transmissivity.

For prescribed lights the value of K shall be 0.8, corresponding to a meteorological visibility of approximately 13 nautical miles.

—— Means corresponding Inland Annex differs, see pages 74-84

(b) A selection of figures derived from the formula is given in the following table:

Range of visibility (luminous range) of light in nautical miles	Luminous intensity of light in candelas for K = 0.8
D	**I**
1	0.9
2	4.3
3	12
4	27
5	52
6	94

Note: The maximum luminous intensity of navigation lights should be limited to avoid undue glare. This shall not be achieved by a variable control of the luminous intensity.

9. Horizontal sectors

(a) (i) In the forward direction, sidelights as fitted on the vessel shall show the minimum required intensities. The intensities shall decrease to reach practical cut-off between 1 degree and 3 degrees outside the prescribed sectors.

(ii) For sternlights and masthead lights and at 22.5 degrees abaft the beam for sidelights, the minimum required intensities shall be maintained over the arc of the horizon up to 5 degrees within the limits of the sectors prescribed in Rule 21. From 5 degrees within the prescribed sectors the intensity may decrease by 50 percent up to the prescribed limits; it shall decrease steadily to reach practical cut-off at not more than 5 degrees outside the prescribed sectors.

(b) All-round lights shall be so located as not to be obscured by masts, topmasts or structures within angular sectors of more than 6 degrees, except anchor lights prescribed in Rule 30, which need not be placed at an impracticable height above the hull.

10. Vertical sectors

(a) The vertical sectors of electric lights as fitted, with the exception of lights on sailing vessels underway shall ensure that:

(i) at least the required minimum intensity is maintained at all angles from 5 degrees above to 5 degrees below the horizontal.

(ii) at least 60 percent of the required minimum intensity is maintained from 7.5 degrees above to 7.5 degrees below the horizontal.

(b) In the case of sailing vessels underway the vertical sectors of electric lights as fitted shall ensure that:

(i) at least the required minimum intensity is maintained at all angles from 5 degrees above to 5 degrees below the horizontal;

(ii) at least 50 percent of the required minimum intensity is maintained from 25 degrees above to 25 degrees below the horizontal.

(c) In the case of lights other than electric these specifications shall be met as closely as possible.

11. Intensity of non-electric lights
Non-electric lights shall so far as practicable comply with the minimum intensities, as specified in the Table given in Section 8 of this Annex.

12. Maneuvering light
Notwithstanding the provisions of paragraph 2(f) of this Annex the maneuvering light described in Rule 34(b) shall be placed in the same fore and aft vertical plane as the masthead light or lights and, where practicable, at a minimum height of 2 meters vertically above the forward masthead light, provided that it shall be carried not less than 2 meters vertically above or below the after masthead light. On a vessel where only one masthead light is carried the maneuvering light, if fitted, shall be carried where it can best be seen, not less than 2 meters vertically apart from the masthead light.

13. Approval
The construction of lights and shapes and the installation of lights on board the vessel shall be to the satisfaction of the appropriate authority of the State whose flag the vessel is entitled to fly.

ANNEX II
Additional Signals for Fishing Vessels Fishing in Close Proximity

1. General
The lights mentioned herein shall, if exhibited in pursuance of Rule 26(d), be placed where they can best be seen. They shall be at least 0.9 meter apart but at a lower level than lights prescribed in Rule 26(b)(i) and (c)(i). The lights shall be visible all around the horizon at a distance of at least 1 mile but at a lesser distance than the lights prescribed by these Rules for fishing vessels.

2. Signals for trawlers

(a) Vessels when engaged in trawling, whether using demersal or pelagic gear, may exhibit:

(i) when shooting their nets: two white lights in a vertical line;

(ii) when hauling their nets: one white light over one red light in a vertical line,

(iii) when the net has come fast upon an obstruction: two red lights in a vertical line.

(b) Each vessel engaged in pair trawling may exhibit:

(i) by night, a searchlight directed forward and in the direction of the other vessel of the pair;

(ii) when shooting or hauling their nets or when their nets have come fast upon an obstruction, the lights prescribed in 2(a) above.

3. Signals for purse seiners

Vessels engaged in fishing with purse seine gear may exhibit two yellow lights in a vertical line. These lights shall flash alternately every second and with equal light and occultation duration. These lights may be exhibited only when the vessel is hampered by its fishing gear.

ANNEX III
Technical Details of Sound Signal Appliances

1. Whistles

(a) <u>Frequencies and range of audibility</u>. The fundamental frequency of the signal shall lie within the range 70-700 Hz. The range of audibility of the signal from a whistle shall be determined by those frequencies, which may include the fundamental and/or one or more higher frequencies, which lie within the range 180-700 Hz (+/- 1 percent) and which provide the sound pressure levels specified in paragraph 1(c) below.

(b) <u>Limits of fundamental frequencies</u>. To ensure a wide variety of whistle characteristics, the fundamental frequency of a whistle shall be between the following limits:

(i) 70-200 Hz, for a vessel 200 meters or more in length;

(ii) 130-350 Hz, for a vessel 75 meters but less than 200 meters in length;

(iii) 250-700 Hz, for a vessel less than 75 meters in length.

(c) <u>Sound signal intensity and range of audibility</u>. A whistle fitted in a vessel shall provide, in the direction of maximum intensity of the whistle and at a distance of 1 meter from it, a sound pressure level in at least one 1/3-octave band within the range of frequencies 180-700 Hz (+/- 1 percent) of not less than the appropriate figure given in the table below.

━━━━━━━━━━ Means corresponding Inland Annex differs, see pages 74-84

Length of vessel in meters	1/3-octave band level at 1 meter in dB referred to 2x10⁻⁵ N/m²	Audibility range in nautical miles
200 or more	143	2
75 but less than 200	138	1.5
20 but less than 75	130	1
Less than 20	120	0.5

The range of audibility in the table above is for information and is approximately the range at which a whistle may be heard on its forward axis with 90 percent probability in conditions of still air on board a vessel having average background noise level at the listening posts (taken to be 68 dB in the octave band centered on 250 Hz and 63 dB in the octave band centered on 500 Hz).

In practice the range at which a whistle may be heard is extremely variable and depends critically on weather conditions; the values given can be regarded as typical but under conditions of strong wind or high ambient noise level at the listening post the range may be much reduced.

(d) Directional properties. The sound pressure level of a directional whistle shall be not more than 4 dB below the prescribed sound pressure level on the axis at any direction in the horizontal plane within +/- 45 degrees of the axis. The sound pressure level at any other direction in the horizontal plane shall be not more than 10 dB below the prescribed sound pressure level on the axis, so that the range in any direction will be at least half the range on the forward axis. The sound pressure level shall be measured in that one-third octave band which determines the audibility range.

(e) Positioning of whistles. When a directional whistle is to be used as the only whistle on a vessel, it shall be installed with its maximum intensity directed straight ahead.

A whistle shall be placed as high as practicable on a vessel, in order to reduce interception of the emitted sound by obstructions and also to minimize hearing damage risk to personnel. The sound pressure level of the vessel's own signal at listening posts shall not exceed 110 dB (A) and so far as practicable should not exceed 100 dB (A).

(f) Fitting of more than one whistle. If whistles are fitted at a distance apart of more than 100 meters, it shall be so arranged that they are not sounded simultaneously.

(g)Combined whistle systems. If due to the presence of obstructions the sound field of a single whistle or of one of the whistles referred to in paragraph 1(f) above is likely to have a zone of greatly reduced signal level, it is recommended that a combined whistle system be fitted so as to overcome this reduction. For the purposes of the Rules a combined whistle system is to be regarded as a single whistle. The whistles of a combined system shall be located at a distance apart of not more than 100 meters and arranged to be sounded simultaneously. The frequency of any one whistle shall differ from those of the others by at least 10 Hz.

2. Bell or gong
(a) Intensity of signal. A bell or gong, or other device having similar sound characteristics shall produce a sound pressure level of not less than 110 dB at a distance of 1 meter from it.

(b)Construction. Bells and gongs shall be made of corrosion-resistant material and designed to give a clear tone. The diameter of the mouth of the bell shall be not less than 300 mm for vessels of 20 meters or more in length, and shall be not less than 200 mm for vessels of 12 meters or more but of less than 20 meters in length. Where practicable, a power-driven bell striker is recommended to ensure constant force but manual operation shall be possible. The mass of the striker shall be not less than 3 percent of the mass of the bell.

3. Approval
The construction of sound signal appliances, their performance and their installation on board the vessel shall be to the satisfaction of the appropriate authority of the State whose flag the vessel is entitled to fly.

ANNEX IV
Distress Signals

1. Need of assistance
The following signals, used or exhibited either together or separately, indicate distress and need of assistance:

(a) a gun or other explosive signal fired at intervals of about a minute;
(b) a continuous sounding with any fog-signaling apparatus;
(c) rockets or shells, throwing red stars fired one at a time at short intervals;
(d) a signal made by radiotelegraphy or by any other signaling method
    consisting of the group . . . _ _ _ . . . (SOS) in the Morse Code;
(e) a signal sent by radiotelephony consisting of the spoken word "Mayday";
(f) the International Code Signal of distress indicated by NC;
(g) a signal consisting of a square flag having above or below it a ball or
    anything resembling a ball;

(h) flames on the vessel (as from a burning tar barrel, oil barrel, etc.);
(i) a rocket parachute or a hand flare showing a red light;
(j) a smoke signal giving off orange colored smoke;
(k) slowly and repeatedly raising and lowering arms outstretched to each side;
(l) the radiotelegraph alarm signal;
(m) the radiotelephone alarm signal;
(n) signals transmitted by emergency position-indicating radio beacons (EPIRB).
(o) approved signals transmitted by radiocommunication systems.

See Inland Annex for additional signal.

2. The use or exhibition of any of the foregoing signals except for the purpose of indicating distress and need of assistance and the use of other signals which may be confused with any of the above signals is prohibited.

3. Attention is drawn to the relevant sections of the International Code of Signals, the Merchant Ship Search and Rescue Manual and the following signals:
(a) a piece of orange-colored canvas with either a black square and circle or other appropriate symbol (for identification from the air);
(b) a dye marker.

See Inland Annex V on Pilot Rules

INTERPRETATIVE RULES

AUTHORITY: 33 U.S.C. 180, 30 Stat. 98; 49 CFR 1.46(c)(2); 28 Stat. 647, 33 U.S.C. 258; 49 CFR 1.46(c)(3); Sec. 4233 R.S., 33 U.S.C. 322.

82.1 Purpose.
(a) This part contains the interpretative rules concerning the 72 COLREGS that are adopted by the Coast Guard for the guidance of the public.

82.3 Pushing vessel and vessel being pushed: Composite unit.
Rule 24(b) of the 72 COLREGS states that when a pushing vessel and a vessel being pushed ahead are rigidly connected in a composite unit, they are regarded as a power-driven vessel and must exhibit the lights under Rule 23. A "composite unit" is interpreted to be a pushing vessel that is rigidly connected by mechanical means to a vessel being pushed so they react to sea and swell as one vessel. "Mechanical means" does not include the following: (a) Lines, (b) Hawsers, (c) Wires, (d) Chains. [CGD 76-133, 42 FR 35792, July 11, 1977. Redesignated by CGD 81-017, 46 FR 28154, May 26, 1981]

————— Means corresponding Inland Annex differs, see pages 74-84

COLREGS Demarcation Lines
Contents

80.01 General basis and purpose of demarcation lines.

AUTHORITY: Rule 1, International Regulations for Preventing Collisions at Sea, 1972 (as rectified); E.O. 11964; Pub. L. 95-75, 91 Stat. 308; 14 U.S.C. 2; 49 CFR 1.46(b), unless otherwise noted.

SOURCE: CGD 77-118a, 42 FR 35784, July 11, 1977, unless otherwise noted.

80.01 General basis and purpose of demarcation lines.

(a) The regulations in this part establish the lines of demarcation delineating those waters upon which mariners shall comply with the International Regulations for Preventing Collisions at Sea, 1972 (72 COLREGS) and those waters upon which mariners shall comply with the Inland Navigation Rules.

(b) The waters inside of the lines are Inland Rules Waters. The waters outside the lines are COLREGS Waters. [CGD 82-029, 47 FR 19518, May 6, 1982; CGD 83-003, 48 FR 7442, Feb. 22, 1983]

ATLANTIC COAST
First District

80.105 Calais, ME to Cape Small, ME. The 72 COLREGS shall apply on the harbors, bays and inlets on the east coast of Maine from International Bridge at Calais, ME to the southwesternmost extremity of Bald Head at Cape Small.

80.110 Casco Bay, ME. (a) A line drawn from the southwesternmost extremity of Bald Head at Cape Small to the southeasternmost extremity of Ragged Island; thence to the southern tangent of Jaquish Island thence to Little Mark Island Monument Light; thence to the northernmost extremity of Jewell Island.

(b) A line drawn from the tower on Jewell Island charted in approximate position latitude 43° 40.6' N longitude 70° 05.9'W to the northeasternmost extremity of Outer Green Island.

(c) A line drawn from the southwesternmost extremity of Outer Green Island to Ram Island Ledge Light; thence to Portland Head Light.

80.115 Portland Head, ME to Cape Ann, MA. (a) Except inside lines specifically described in this section, the 72 COLREGS shall apply on the harbors, bays, and inlets on the east coast of Maine, New Hampshire, and Massachusetts from Portland Head to Halibut Point at Cape Ann.

(b) A line drawn from the southernmost tower on Gerrish Island charted in approximate position latitude 43° 04.0' N longitude 70° 41.2' W to Whaleback Light; thence to Jeffrey Point Light 2A; thence to the northeasternmost extremity of Frost Point.

(c) A line drawn from the northernmost extremity of Farm Point to Annisquam Harbor Light.

80.120 Cape Ann, MA to Marblehead Neck, MA. (a) Except inside lines specifically described in this section, the 72 COLREGS shall apply on the harbors, bays and inlets on the east coast of Massachusetts from Halibut Point at Cape Ann to Marblehead Neck.

(b) A line drawn from Gloucester Harbor Breakwater Light to the twin towers charted in approximate position latitude 42° 35.1' N longitude 70° 41.6' W.

(c) A line drawn from the westernmost extremity of Gales Point to the easternmost extremity of House Island; thence to Bakers Island Light; thence to Marblehead Light. [CGD 81-017, 46 FR 28154, May 26, 1981]

80.125 Marblehead Neck, MA to Nahant, MA. The 72 COLREGS apply on the harbors, bays, and inlets on the east coast of Massachusetts from Marblehead Neck to the easternmost tower at Nahant, charted in approximate position latitude 42° 25.4' N, longitude 70° 54.6' W. [CGD 81-017, 46 FR 28154, May 26, 1981]

80.130 Boston Harbor entrance. A line drawn from easternmost tower at Nahant, charted in approximate position latitude 42° 25.4' N, longitude 70° 54.6' W, to Boston Lighted Horn

Buoy "B"; thence to the easternmost radio tower at Hull, charted in approximate position latitude 42° 16.7' N, longitude 70° 52.6'W. [CGD 81-017, 46 FR 28154, May 26, 1981]

80.135 Hull, MA to Race Point, MA.

(a) Except inside lines described in this section, the 72 COLREGS apply on the harbors, bays, and inlets on the east coast of Massachusetts from the easternmost radio tower at Hull, charted in approximate position latitude 42° 16.7'N, longitude 70° 52.6' W, to Race Point on Cape Cod.

(b) A line drawn from Canal Breakwater Light 4 south to the shoreline. [CGD 81-017, 46 FR 28154, May 26, 1981]

80.145 Race Point, MA to Watch Hill, RI. (a) Except inside lines specifically described in this section, the 72 COLREGS shall apply on the sounds, bays, harbors and inlets along the coast of Cape Cod and the southern coasts of Massachusetts and Rhode Island from Race Point to Watch Hill.

(b) A line drawn from Nobska Point Light to Tarpaulin Cove Light on the southeastern side of Naushon Island; thence from the southernmost tangent of Naushon Island to the easternmost extremity of Nashawena Island; thence from the southwestern most extremity of Nashawena Island to the easternmost extremity of Cuttyhunk Island; thence from the southwestern tangent of Cuttyhunk Island to the tower on Gooseberry Neck charted in approximate position latitude 41° 29.1' N, longitude 71° 02.3' W.

(c) A line drawn from Sakonnet Breakwater Light 2 tangent to the southernmost part of Sacnuest Point charted in approximate position latitude 41° 28.5'N, longitude 71° 14.8'W.

(d) An east-west line drawn through Beavertail Light between Brenton Point and the Boston Neck shoreline. [CGD 81-017, 46 FR 28154, May 26, 1981]

80.150 Block Island, RI. The 72 COLREGS shall apply on the harbors of Block Island.

80.155 Watch Hill, RI to Montauk Point, NY.

(a) A line drawn from Watch Hill Light to East Point on Fishers Island.

(b) A line drawn from Race Point to Race Rock Light; thence to Little Gull Island Light thence to East Point on Plum Island.

(c) A line drawn from Plum Island Harbor East Dolphin Light to Plum Island Harbor West Dolphin Light.

(d) A line drawn from Plum Island Light to Orient Point Light; thence to Orient Point.

(e) A line drawn from the lighthouse ruins at the southwestern end of Long Beach Point to Cornelius Point.

(f) A line drawn from Coecles Harbor Entrance Light to Sungic Point.

(g) A line drawn from Nichols Point to Cedar Island Light.

(h) A line drawn from Three Mile Harbor West Breakwater Light to Three Mile Harbor East Breakwater Light.

(i) A line drawn from Montauk West Jetty Light 1 to Montauk East Jetty Light 2. [Redisignated by CGD 87-0086, 52 FR 25218, July 6, 1987]

80.160 Montauk Point, NY to Atlantic Beach, NY. (a) A line drawn from Shinnecock Inlet East Breakwater Light to Shinnecock Inlet West Breakwater Light 1.

(b) A line drawn from Moriches Inlet East Breakwater Light to Moriches Inlet West Breakwater Light.

(c) A line drawn from Fire Island Inlet Breakwater Light 348° true to the southernmost extremity of the spit of land at the western end of Oak Beach.

(d) A line drawn from Jones Inlet Light 322° true across the southwest tangent of the island on the north side of Jones Inlet to the shoreline. [Redesignated CGD 87-0086, 52 FR 25218, July 6, 1987]

80.165 New York Harbor. A line drawn from East Rockaway Inlet Breakwater Light to Sandy Hook Light.

80.170 Sandy Hook, NJ to Tom's River, NJ (a) A line drawn from Shark River Inlet North Breakwater Light 2 to Shark River Inlet South Breakwater Light 1.

(b) A line drawn from Manasquan Inlet North Breakwater Light 4 to Manasquan Inlet South Breakwater Light 3.

(c) A line drawn from Barnegat Inlet North Breakwater Light 4A to the seaward extremity of the submerged Barnegat Inlet South Breakwater; thence along the submerged breakwater to the shoreline.

Fifth District

80.501 Tom's River NJ to Cape May, NJ. (a) A line drawn from the seaward tangent of Long Beach Island to the seaward tangent to Pullen Island across Beach Haven and Little Egg Inlets.

(b) A line drawn from the seaward tangent of Pullen Island to the seaward tangent of Brigantine Island across Brigantine Inlet.

(c) A line drawn from the seaward extremity of Absecon Inlet North Jetty to Atlantic City light.

(d) A line drawn from the southernmost point of Longport at latitude 30° 18.2' N, longitude 75° 32.2' W to the northeasternmost point of Ocean City at latitude 39° 17.6' N, longitude 74° 33.1' W across the Great Egg Harbor Inlet.

(e) A line drawn parallel with the general trend of highwater shoreline across Corson Inlet.

(f) A line formed by the centerline of the Townsend Inlet Highway Bridge.

(g) A line formed by the shoreline of Seven Mile Beach and Hereford Inlet Light.

(h) A line drawn from Cape May Inlet West Jetty Light. [CGD 87-0086, 52 FR 25218, July 6, 1987]

80.503 Delaware Bay. A line drawn from Cape May Light to Harbor of Refuge Light; thence to the northernmost extremity of Cape Henlopen.

80.505 Cape Henlopen, DE to Cape Charles, VA. (a) A line drawn from the seaward extremity of Indian River Inlet North Jetty to Indian River Inlet South Jetty Light.

(b) A line drawn from Ocean City Inlet Light 6 225° true across Ocean City Inlet to the submerged south breakwater.

(c) A line drawn from Assateague Beach Tower Light to the tower charted at latitude 37° 52.6' N, longitude 75° 26.7' W.

(d) A line formed by the range of Wachapreague Inlet Light 3 and Parramore Beach lookout Tower drawn across Wachapreague Inlet.

(e) A line drawn from the lookout tower charted on the northern end of Hog Island to the seaward tangent of Parramore Beach.

(f) A line drawn 207° true from the lookout tower charted on the southern end of Hog Island across Great Machipongo Inlet.

(g) A line formed by the range of the two cupolas charted on the southern end of Cobb Island drawn across the Sand Shoal Inlet.

(h) Except as provided elsewhere in this section from Cape Henlopen to Cape Charles, lines drawn parallel with the general trend of the highwater shoreline across the entrances to small bays and inlets.

80.510 Chesapeake Bay Entrance, VA. A line drawn from Cape Charles Light to Cape Henry Light.

80.515 Cape Henry, VA to Cape Hatteras, NC. (a) A line drawn from Rudee Inlet Jetty Light 2 to Rudee Inlet Jetty Light 1.

(b) A line formed by the centerline of the highway bridge across Oregon Inlet.

80.520 Cape Hatteras, NC to Cape Lookout, NC. (a) A line drawn from Hatteras Inlet Lookout Tower (30° 11.8' N, 75° 44.9' W) 255° true to the eastern end of Ocracoke Island.

(b) A line drawn from the westernmost extremity of Ocracoke Island at latitude 35° 04.0' N, longitude 76° 00.8' W to the northeastern extremity of Portsmouth Island at latitude 35° 03.7' N, longitude 76° 02.3' W.

(c) A line drawn across Drum Inlet parallel with the general trend of the highwater shoreline.

80.525 Cape Lookout, NC to Cape Fear, NC. (a) A line drawn from Cape Lookout Light to the seaward tangent of the southeastern end of Shackleford Banks.

(b) A line drawn from Morehead City Channel Range Front Light to the seaward extremity of the Beaufort Inlet west jetty.

(c) A line drawn from the southernmost extremity of Bogue Banks at latitude 34° 38.7' N, longitude 77° 06.0' W across Bogue Inlet to the northernmost extremity of Bear Beach at latitude 34° 38.5' N, longitude 77° 07.1' W.

(d) A line drawn from the tower charted in approximate position latitude 34° 31.5' N, longitude 77° 20.8' W to the seaward tangent of the shoreline on the northeast side of New River Inlet.

(e) A line drawn across New Topsail Inlet between the closest extremities of the shore on either side of the inlet from latitude 34° 20.8' N, longitude 77° 39.2' W to latitude 34° 20.6' N, longitude 77° 39.6' W.

(f) A line drawn from the seaward extremity of the jetty on the northeast side of Masonboro Inlet to the seaward extremity of the jetty on the southeast side of the Inlet.

(g) Except as provided elsewhere in this section from Cape Lookout to Cape Fear, lines drawn parallel with the general trend of the highwater shoreline across the entrance of small bays and inlets.

80.530 Cape Fear, NC to Little River Inlet, NC. (a) A line drawn from the abandoned lighthouse charted in approximate position latitude 33° 52.4' N longitude 78° 00.1 W across the Cape Fear River Entrance to Oak Island Light.

(b) Except as provided elsewhere in this section from Cape Fear to Little River Inlet, lines drawn parallel with the general trend of the highwater shoreline across the entrance to small inlets.

Seventh District

80.703 Little River Inlet, SC to Cape Romain, SC. (a) A line drawn from the westernmost extremity of the sand spit on Bird Island to the easternmost extremity of Waties Island across Little River Inlet.

(b) From Little River Inlet, a line drawn parallel with the general trend of the highwater shoreline across Hog Inlet; thence a line drawn from Murrels Inlet Light 2 to Murrels Inlet Light 1; thence a line drawn parallel with the general trend of the highwater shoreline across Midway Inlet, Pawleys Inlet and North Inlet.

(c) A line drawn from the charted position of Winyah Bay North Jetty End Buoy 2N south to the Winyah Bay South Jetty.

(d) A line drawn from Santee Point to the seaward tangent of Cedar Island.

(e) A line drawn from Cedar Island Point west to Murphy Island.

(f) A north-south line (longitude 79° 20.3' W) drawn from Murphy Island to the northernmost extremity of Cape Island Point.

80.707 Cape Romain, SC to Sullivans Island, SC. (a) A line drawn frrom the western extremity of Cape Romain 292° true to Racoon Key on the west side of Racoon Creek.

(b) A line drawn from the westernmost extremity of Sandy Point across Bull Bay to the northernmost extremity of Northeast Point.

(c) A line drawn from the southernmost extremity of Bull Island to the easternmost extremity of Capers Island.

(d) A line formed by the overhead power cable from Capers Island to Dewees Island.

(e) A line formed by the overhead power cable from Dewees Island to Isle of Palms.

(f) A line formed by the centerline of the highway bridge between Isle of Palms and Sullivans Island over Breach Inlet.

80.710 Charleston Harbor, SC. (a) A line formed by the submerged north jetty from the shore to the west end of the north jetty.

(b) A line drawn from across the seaward extremity of the Charleston Harbor Jetties.

(c) A line drawn from the west end of the South

Jetty across the South entrance to Charleston Harbor to shore on a line formed by the submerged south jetty.

80.712 Morris Island, SC to Hilton Head Island, SC. (a) A line drawn from the easternmost tip of Folley Island to the abandoned lighthouse tower on the northside of Lighthouse Inlet; thence west to the shoreline of Morris Island.

(b) A straight line drawn from the seaward tangent of Folly Island through Folly River Daybeacon 10 across Stono River to the shoreline of Sandy Point.

(c) A line drawn from the southernmost extremity of Seabrook Island 257° true across the North Edisto River Entrance to the shore of Botany Bay Island.

(d) A line drawn from the microwave antenna tower on Edisto Beach charted in approximate position latitude 32° 29.3' N, longitude 80° 19.2' W across St. Helena Sound to the abandoned lighthouse tower on Hunting Island.

(e) A line formed by the centerline of the highway bridge between Hunting Island and Fripp Island.

(f) A line following the general trend of the seaward highwater shoreline across Cabretta Inlet.

80.715 Savannah River. A line drawn from the southernmost tank on Hilton Head Island charted in approximate position latitude 32° 06.7'N, longitude 80° 49.3' W to Bloody Point Range Rear Light; thence to Tybee (Range Rear) Light.

80.717 Tybee Island, GA to St. Simons Island, GA. (a) A line drawn from the southernmost extremity of Savannah Beach on Tybee Island 255° true across Tybee Inlet to the shore of Little Tybee Island south of the entrance to Buck Hammock Creek.

(b) A straight line drawn from the northernmost extremity of Wassaw Island 031° true through Tybee River Daybeacon 1 to the shore of Little Tybee Island.

(c) A line drawn approximately parallel with the general trend of the highwater shorelines from the seaward tangent of Wassau Island to the seaward tangent of Bradley Point on Ossabaw Island.

(d) A north-south line (longitude 81° 08.4'W) drawn from the southernmost extremity of Ossabaw Island to St. Catherines Island .

(e) A north-south line (longitude 81°10.6' W) drawn from the southernmost extremity of St.

Catherines Island to Northeast Point on Blackbeard Island.

(f) A line following the general trend of the seaward highwater shoreline across Cabretta Inlet.

(g) A north-south line (longitude 81° 16.9' W) drawn from the southwesternmost point on Sapelo Island to Wolf Island.

(h) A north-south line (longitude 81°17.1' W) drawn from the southeasternmost point of Wolf Island to the northeasternmost point on Little St. Simons Island.

(i) A line drawn from the northeasternmost extremity of Sea Island 045° true to Little St. Simons Island.

(j) An east-west line from the southernmost extremity of Sea Island across Goulds Inlet to St. Simons Island.

80.720 St. Simons Island, GA to Amelia Island, FL. (a) A line drawn from St. Simons Light to the northernmost tank on Jekyll Island charted in approximate position latitude 31°05.9' N longitude 81° 24.5' W.

(b) A line drawn from the southernmost tank on Jekyll Island charted in approximate position latitude 31° 01.6' N longitude 81° 25.2' W to coordinate latitude 30° 59.4' N, longitude 81° 23.7' W (0.5 nautical mile east of the charted position of St. Andrew Sound Lighted Buoy 32); thence to the abandoned lighthouse tower on the north end of Little Cumberland Island charted in approximate position lat. 30° 58.5' N, lon. 81° 24.8' W.

(c) A line drawn across the seaward extremity of the St. Marys Entrance Jetties.

80.723 Amelia Island, FL to Cape Canaveral, FL. (a) A line drawn from the southernmost extremity of Amelia Island to the northeasternmost extremity of Little Talbot Island.

(b) A line formed by the centerline of the highway bridge from Little Talbot Island to Fort George Island.

(c) A line drawn across the seaward extremity of the St. Johns River Entrance Jetties.

(d) A line drawn across the seaward extremity of the St. Augustine Inlet Jetties.

(e) A line formed by the centerline of the highway bridge over Matanzas Inlet.

(f) A line drawn across the seaward extremity of the Ponce de Leon Inlet Jetties.

80.727 Cape Canaveral, FL to Miami Beach, FL. (a) A line drawn across the seaward extremity of the Port Canaveral Entrance Channel Jetties.

(b) A line drawn across the seaward extremity of the Sebastian Inlet Jetties.

(c) A line drawn across the seaward extremity of the Fort Pierce Inlet Jetties.

(d) A north-south line (longitude 80° 09.7' W) drawn across St. Lucie Inlet.

(e) A line drawn from the seaward extremity of Jupiter Inlet North Jetty to the northeast extremity of the concrete apron on the south side of Jupiter Inlet.

(f) A line drawn across the seaward extremity of the Lake Worth Inlet Jetties.

(g) A line drawn across the seaward extremity of the Boynton Inlet Jetties.

(h) A line drawn from Boca Raton Inlet North Jetty Light 2 to Boca Raton Inlet South Jetty Light 1.

(i) A line drawn from Hillsboro Inlet Light to Hillsboro Inlet Entrance Light 2; thence to Hillsboro Inlet Entrance Light 1; thence west to the shoreline.

(j) A line drawn across the seaward extremity of the Port Everglades Entrance Jetties.

(k) A line formed by the centerline of the highway bridge over Bakers Haulover Inlet.

80.730 Miami Harbor, FL. A line drawn across the seaward extremity of the Miami Harbor Government Cut Jetties.

80.735 Miami, FL to Long Key, FL. (a) A line drawn from the southernmost extremity of Fisher Island 212° true to the point latitude 25° 45.0' N longitude 80° 08.6' W on Virginia Key.

(b) A line formed by the centerline of the highway bridge between Virginia Key and Key Biscayne.

(c) A line drawn from Cape Florida Light to the northernmost extremity on Soldier Key.

(d) A line drawn from the southernmost extremity on Soldier Key to the northernmost extremity of the Ragged Keys.

(e) A line drawn from the Ragged Keys to the southernmost extremity of Angelfish Key following the general trend of the seaward shoreline.

(f) A line drawn on the centerline of the Overseas Highway (U.S. 1) and bridges from latitude 25° 19.3' N longitude 80° 16.0' W at Little Angelfish Creek to the radar dome charted on Long Key at approximate position latitude 24°

49.3' N longitude 80° 49.2' W [CGD 81-017, 46 FR 28154, May 26, 1981]

PUERTO RICO AND VIRGIN ISLANDS

Seventh District

80.738 Puerto Rico and Virgin Islands. (a) Except inside lines specifically described in this section, the 72 COLREGS shall apply on all other bays, harbors and lagoons of Puerto Rico and the U.S. Virgin Islands.

(b) A line drawn from Puerto San Juan Light to Cabras Light across the entrance of San Juan Harbor.

GULF COAST

Seventh District

80.740 Long Key, FL to Cape Sable, FL. A line drawn from the microwave tower charted on Long Key at approximate position latitude 24° 48.8' N longitude 80° 49.6' W to Long Key Light 1; thence to Arsenic Bank Light 2; thence to Sprigger Bank Light 5; thence to Schooner Bank Light 6; thence to Oxfoot Bank Light 10; thence to East Cape Light 2; thence through East Cape Daybeacon 1A to the shoreline at East Cape.

80.745 Cape Sable, FL to Cape Romano, FL. (a) A line drawn following the general trend of the mainland, highwater shoreline from Cape Sable at East Cape to Little Shark River Light 1; thence to westernmost extremity of Shark Point; thence following the general trend of the mainland, highwater shoreline crossing the entrances of Harney River, Broad Creek, Broad River, Rodgers River First Bay, Chatham River, Huston River, to the shoreline at coordinate latitude 25° 41.8' N longitude 81° 17.9' W.

(b) The 72 COLREGS shall apply to the waters surrounding the Ten Thousand Islands and the bays, creeks, inlets and rivers between Chatham Bend and Marco Island except inside lines specifically described in this part.

(c) A north-south line drawn at longitude 81° 20.2' W across the entrance to Lopez River.

(d) A line drawn across the entrance to Turner River parallel to the general trend of the shoreline.

(e) A line formed by the centerline of Highway 92 Bridge at Goodland.

80.748 Cape Romano, FL to Sanibel Island, FL. (a) A line drawn across Big Marco Pass parallel to the general trend of the seaward, highwater shoreline.

(b) A line drawn from the northwesternmost extremity of Coconut Island 000° T across Capri Pass.

(c) Lines drawn across Hurricane and Little Marco Passes parallel to the general trend of the seaward, highwater shoreline.

(d) A line from the seaward extremity of Gordon Pass South Jetty 014° true to the shoreline at approximate coordinate latitude 26° 05.7' N longitude 81° 48.1' W.

(e) A line drawn across the seaward extremity of Doctors Pass Jetties.

(f) Lines drawn across Wiggins, Big Hickory, New, and Big Carlos Passes parallel to the general trend of the seaward highwater shoreland.

(g) A straight line drawn from Sanibel Island Light through Matanzas Pass Channel Light 2 to the shore of Estero Island. [CGD 81-017, 46 FR 28154, May 26, 1981]

80.750 Sanibel Island, FL to St. Petersburg, FL. (a) A line formed by the centerline of the highway bridge over Blind Pass, between Captiva Island and Sanibel Island, and lines drawn across Redfish and Captiva Passes parallel to the general trend of the seaward, highwater shorelines.

(b) A line drawn from La Costa Test Pile North Light to Port Boca Grande Light.

(c) Lines drawn across Gasparilla and Stump Passes parallel to the general trend of the seaward, highwater shorelines.

(d) A line across the seaward extremity of Venice Inlet Jetties.

(e) A line drawn across Midnight Pass parallel to the general trend of the seaward, highwater shoreline.

(f) A line drawn from Big Sarasota Pass Light 14 to the southernmost extremity of Lido Key.

(g) A line drawn across New Pass tangent to the seaward, highwater shoreline of Longboat Key.

(h) A line drawn across Longboat Pass parallel to the seaward, highwater shoreline.

(i) A line drawn from the northwesternmost extremity of Bean Point to the southeasternmost extremity of Egmont Key.

(j) A straight line drawn from Egmont Key Light through Egmont Channel Range Rear Light to the shoreline on Mullet Key.

(k) A line drawn from the northernmost extremity of Mullet Key across Bunces Pass and South Channel to Pass-a-Grille Channel Light 8; thence to Pass-a-Grille Channel Daybeacon 9; thence to the southwesternmost extremity of Long Key. [CGD 81-017, 46 FR 28154, May 26, 1981]

80.753 St. Petersburg, FL to the Anclote, FL. (a) A line drawn across Blind Pass, between Treasure Island and Long Key, parallel with the general trend of the seaward, highwater shoreline.

(b) Lines formed by the centerline of the highway bridges over Johns and Clearwater Passes.

(c) A line drawn across Dunedin and Hurricane Passes parallel with the general trend of the seaward, highwater shoreline.

(d) A line drawn from the northernmost extremity of Honeymoon Island to Anclote Anchorage South Entrance Light 7; thence to Anclote Key 28° 10.0' N, 82° 50.6' W; thence a straight line through Anclote River Cut B Range Rear Light to the shoreline.

80.755 Anclote, FL to the Suncoast Keys, FL. (a) Except inside lines specifically described in this section, the 72 COLREGS shall apply on the bays, bayous, creeks, marinas, and rivers from Anclote to the Suncoast Keys.

(b) A north-south line drawn at longitude 82° 38.3' W across the Chassahowitzka River Entrance.

80.757 Suncoast Keys, FL to Horseshoe Point, FL. (a) Except inside lines specifically described in this section, the 72 COLREGS shall apply on the bays, bayous, creeks, and marinas from Suncoast Keys to Horseshoe Point.

(b) A line formed by the centerline of Highway 44 Bridge over the Salt River.

(c) A north-south line drawn through Crystal River Entrance Daybeacon 25 across the river entrance.

(d) A north-south line drawn through the Cross Florida Barge Canal Daybeacon 48 across the canal.

(e) A north-south line drawn through Withlacoochee River Daybeacon 40 across the river.

(f) A line drawn from the westernmost extremity of South Point north to the shoreline across the Waccasassa River Entrance.

(g) A line drawn from position latitude 29° 16.6' N longitude 83° 06.7' W 300° true to the shoreline of Hog Island.

(h) A north-south line drawn through Suwannee

RiverWadley Pass Channel Daybeacons 30 and 31 across the Suwanee River.

80.760 Horseshoe Point, FL to Rock Islands, FL. (a) Except inside lines specifically described provided in this section, the 72 COLREGS shall apply on the bays, bayous, creeks, marinas, and rivers from Horseshoe Point to the Rock Islands.

(b) A north-south line drawn through Steinhatchee River Light 21.

(c) A line drawn from Fenholloway River Approach Light FR east across the entrance to Fenholloway River.

Eighth District

80.805 Rock Island, FL to Cape San Blas, FL. (a) A north-south line drawn from the Econfina River Light to the opposite shore.

(b) A line drawn from Gamble Point Light to the southernmost extremity of Cabell Point.

(c) A line drawn from St. Marks (Range Rear) Light to St. Marks Channel Light 11; thence to the southernmost extremity of Live Oak Point; thence in a straight line through Shell Point Light to the southernmost extremity of Ochlockonee Point; thence to Bald Point along longitude 84° 20.5' W.

(d) A line drawn from the south shore of Southwest Cape at longitude 84° 22.7' W to Dog Island Reef East Light 1; thence to Turkey Point Light 2; thence to the easternmost extremity of Dog Island.

(e) A line drawn from the westernmost extremity of Dog Island to the easternmost extremity of St. George Island.

(f) A line drawn across the seaward extremity of the St. George Island Channel Jetties.

(g) A line drawn from the northwesternmost extremity of Sand Island to West Pass Light 7.

(h) A line drawn from the westernmost extremity of St. Vincent Island to the southeast, highwater shoreline of Indian Peninsula at longitude 85° 13.5'W.

80.810 Cape San Blas, FL to Perdido Bay, FL. (a) A line drawn from St. Joseph Bay Entrance Range A Rear Light through St. Joseph Bay Entrance Range B Front Light to St. Joseph Point.

(b) A line drawn across the mouth of Salt Creek as an extension of the general trend of the shoreline to continue across the inlet to St. Andrews sound in the middle of Crooked Island.

(c) A line drawn from the northernmost extremity of Crooked Island 000° T to the mainland.

(d) A line drawn from the easternmost extremity of Shell Island 120° true to the shoreline across the east entrance to St. Andrews Bay.

(e) A line drawn between the seaward end of the St. Andrews Bay Entrance Jetties.

(f) A line drawn between the seaward end of the Choctawatchee Bay Entrance Jetties.

(g) An east-west line drawn from Fort McRee Leading Light across the Pensacola Bay Entrance along latitude 30° 19.5' N.

(h) A line drawn between the seaward end of the Perdido Pass Jetties.

80.815 Mobile Bay, AL to the Chandeleur Islands, LA. (a) A line drawn across the inlets to Little Lagoon as an extension of the general trend of the shoreline.

(b) A line drawn from Mobile Point Light to Dauphin Island Channel Light No. 1 to the eastern corner of Fort Gaines at Pelican Point.

(c) A line drawn from the westernmost extremity of Dauphin Island to the easternmost extremity of Petit Bois Island.

(d) A line drawn from Horn Island Pass Entrance Range Front Light on Petit Bois Island to the easternmost extremity of Horn Island.

(e) An east-west line (latitude 30° 14.7' N) drawn between the westernmost extremity of Horn Island to the easternmost extremity of Ship Island.

(f) A curved line drawn following the general trend of the seaward, highwater shoreline of Ship Island.

(g) A line drawn from Ship Island Light; to Chandeleur Light; thence in a curved line following the general trend of the seaward, highwater shorelines of the Chandeleur Islands to the island at latitude 29° 44.1' N, longitude 88° 53.0' W; thence to lat. 29° 26.5' N, lon. 88° 55.6' W.

80.825 Mississippi Passes, LA. (a) A line drawn from latitude 29° 26.5' N, longitude 88° 55.6' W to latitude 29° 10.6' N, longtitude 88° 59.8' W; thence to latitude 29° 03.5' N, longitude 89° 03.7' W; thence to lat. 28° 58.8'N, lon. 89° 04.3' W.

(b) A line drawn from latitude 28° 58.8' N, longitude 89° 04.3' W; to latitude 28° 57.3' N, longitude 89° 05.3' W; thence to latitude 28° 56.95' N, longitude 89° 05.6' W; thence to latitude 29° 00.4' N, longitude 89° 09.8' W; thence following the general trend of the seaward highwater shoreline in a northwesterly direction to latitude 29° 03.4' N, longitude 89° 13.0' W; thence west to latitude 29° 03.5'N, longitude 89° 15.5'W; thence following the general trend of the seaward highward shoreline in a southwesterly direction to latitude 28° 57.7' N, longitude 89° 22.3'W.

(c) A line drawn from latitude 28° 57.7' N, longitude 89° 22.3' W; to latitude 28° 51.4' N, longitude 89° 24.5' W; thence to latitude 28° 51.5' N, longitude 89° 27.1' W; thence to latitude 28° 52.65' N, longitude 89° 27.1' W; thence to the seaward extremity of the Southwest Pass West Jetty located at latitude 28° 54.5' N, longitude 89° 26.1' W.

(d) A line drawn from Mississippi River South Pass East Jetty Light 4 to Mississippi River South Pass West Jetty Light; thence following the general trend of the seaward highwater shoreline in a northwesterly direction to coordinate latitude 29° 03.4' N, longitude 89° 13.0' W; thence west to coordinate latitude 29° 03.5' N, longitude 89° 15.5' W; thence following the general trend of the seaward highwater shoreline in a southwesterly direction to Mississippi River Southwest Pass Entrance Light.

(e) A line drawn from Mississippi River Southwest Pass Entrance Light; thence to the seaward extremity of the Southwest Pass West Jetty located at coordinate latitude 28° 54.5' N, longitude 89° 26.1' W.

80.830 Mississippi Passes, LA to Point Au Fer, LA. (a) A line drawn from the seaward extremity of the Southwest Pass West Jetty located at coordinate latitude 28° 54.5' N longitude 89° 26.1' W; thence following the general trend of the seaward, highwater jetty and shoreline in a north, northeasterly direction to Old Tower latitude 28° 58.8' N longitude 89° 23.3' W; thence to West Bay Light; thence to coordinate latitude 29° 05.2' N longitude 89° 24.3' W; thence a curved line following the general trend of the highwater shoreline to Point Au Fer Island except as otherwise described in this section.

(b) A line drawn across the seaward extremity of the Empire Waterway (Bayou Fontanelle) entrance jetties.

(c) An east-west line drawn from the westernmost extremity of Grand Terre Islands in the direction of 194° true to the Grand Isle Fishing Jetty Light.

(d) A line drawn between the seaward extremity of the Belle Pass Jetties.

(e) A line drawn from the westernmost extremity of the Timbalier Island to the easternmost extremity of Isles Dernieres.

(f) A north-south line drawn from Caillou Bay Light 13 across Caillou Boca.

(g) A line drawn 107° true from Caillou Bay Boat Landing Light across the entrances to Grand Bayou du Large and Bayou Grand Caillou.

(h) A line drawn on an axis of 103° true through Taylors Bayou Entrance Light 2 across the entrances to Jack Stout Bayou, Taylors Bayou, Pelican Pass and Bayou de West.

80.835 Point Au Fer, LA to Calcasieu Pass, LA. (a) A line drawn from Point Au Fer to Atchafalaya Channel Light 34; thence Point Au Fer Reef Light 33; Atchafalaya Bay Pipeline Light D latitude 29° 25.0' N, longitude 91° 31.7' W; thence Atchafalaya Bay Light 1 latitude 29° 25.3' N, longitude 91° 35.8' W; thence South Point.

(b) Lines following the general trend of the highwater shoreline drawn across the bayou canal inlets from the Gulf of Mexico between South Point and Calcasieu Pass except as otherwise described in this section.

(c) A line drawn on an axis of 140° true through Southwest Pass-Vermillion Bay Light 4 across Southwest Pass.

(d) A line drawn across the seaward extremity of the Freshwater Bayou Canal Entrance Jetties.

(e) A line drawn from Mermentau Channel East Jetty Light 6 to Mermentau Channel West Jetty Light 7.

(f) A line drawn from the radio tower charted in approximate position latitude 29° 45.7' N, longitude 93.06.3' W 115° true across Mermentau Pass.

(g) A line drawn across the seaward extremity of the Calcasieu Pass Jetties.

80.840 Sabine Pass, TX to Galveston, TX. (a) A line drawn from the Sabine Pass East Jetty Light to the seaward end of the Sabine Pass West Jetty.

(b) A line drawn across the small boat passes through the Sabine Pass East and West Jetties.

(c) A line formed by the centerline of the highway bridge over Rollover Pass at Gilchrist.

80.845 Galveston, TX to Freeport, TX. (a) A line drawn from Galveston North Jetty Light 6A to Galveston South Jetty Light 5A.

(b) A line formed by the centerline of the highway bridge over San Luis Pass.

(c) Lines formed by the centerlines of the highway bridges over the inlets to Christmas Bay

(Cedar Cut) and Drum Bay.

(d) A line drawn from the seaward extremity of the Freeport North Jetty to Freeport Entrance Light 6; thence Freeport Entrance Light 7; thence the seaward extremity of Freeport South Jetty.

80.850 Brazos River, TX to the Rio Grande, TX. (a) Except as otherwise described in this section lines drawn continuing the general trend of the seaward, highwater shorelines across the inlets to Brazos River Diversion Channel, San Bernard River, Cedar Lakes, Brown Cedar Cut, Colorado River, Matagorda Bay, Cedar Bayou, Corpus Christi Bay, and Laguna Madre.

(b) A line drawn across the seaward extremity of Matagorda Ship Channel North Jetties.

(c) A line drawn from the seaward tangent of Matagorda Peninsula at Decros Point to Matagorda Light.

(d) A line drawn across the seaward extremity of the Aransas Pass Jetties.

(e) A line drawn across the seaward extremity of the Port Mansfield Entrance Jetties.

(f) A line drawn across the seaward extremity of the Brazos Santiago Pass Jetties.

PACIFIC COAST

Eleventh District

80.1105 Santa Catalina Island, CA. The 72 COLREGS shall apply to the harbors on Santa Catalina Island.

80.1110 San Diego Harbor, CA. A line drawn from Zuniga Jetty Light "V" to Zuniga Jetty Light "Z"; thence to Point Loma Light.

80.1115 Mission Bay, CA. A line drawn from Mission Bay South Jetty Light 2 to Mission Bay North Jetty Light 1.

80.1120 Oceanside Harbor, CA. A line drawn from Oceanside South Jetty Light 4 to Oceanside Breakwater Light 3.

80.1125 Dana Point Harbor, CA. A line drawn from Dana Point Jetty Light 6 to Dana Point Breakwater Light 5.

80.1130 Newport Bay, CA. A line drawn from Newport Bay East Jetty Light 4 to Newport Bay West Jetty Light 3.

80.1135 San Pedro Bay - Anaheim Bay, CA. (a) A line drawn across the seaward extremities of the Anaheim Bay Entrance East Jetties; thence to Long Beach Breakwater East End Light 1.

(b) A line drawn from Long Beach Channel Entrance Light 2 to Long Beach Light.

(c) A line drawn from Los Angeles Main Entrance Channel Light 2 to Los Angeles Light.

80.1140 Redondo Harbor, CA. A line drawn from Redondo Beach East Jetty Light 2 to Redondo Beach West Jetty Light 3.

80.1145 Marina Del Rey, CA. (a) A line drawn from Marina Del Rey Breakwater South Light 1 to Marina Del Rey Light 4.

(b) A line drawn from Marina Del Rey Breakwater North Light 2 to Marina Del Rey Light 3.

(c) A line drawn from Marina Del Rey Light 4 to the seaward extremity of the Ballona Creek South Jetty.

80.1150 Port Hueneme, CA. A line drawn from Port Hueneme East Jetty Light 4 to Port Hueneme West Jetty Light 3.

80.1155 Channel Islands Harbor, CA. (a) A line drawn from Channel Islands Harbor South Jetty Light 2 to Channel Islands Harbor Breakwater South Light 1.

(b) A line drawn from Channel Islands Harbor Breakwater North Light to Channel Islands Harbor North Jetty Light 5.

80.1160 Ventura Marina, CA. A line drawn from Ventura Marina South Jetty Light 6 to Ventura Marina Breakwater South Light 3; thence to Ventura Marina North Jetty Light 7.

80.1165 Santa Barbara Harbor, CA. A line drawn from Santa Barbara Harbor Light 4 to Santa Barbara Harbor Breakwater Light.

Twelfth District

80.1205 San Luis Obispo Bay, CA. A line drawn from the southernmost extremity of Fossil Point to the seaward extremity of Whaler Is. Breakwater.

80.1210 Estero-Morro Bay, CA. A line drawn from the seaward extremity of the Morro Bay East Breakwater to the Morro Bay West Breakwater Light.

80.1215 Monterey Harbor, CA. A line drawn from Monterey Harbor Light 6 to the northern extremity of Monterey Municipal Wharf 2.

80.1220 Moss Landing Harbor, CA. A line drawn from the seaward extremity of the pier located 0.3 mile south of Moss Landing Harbor Entrance to the seaward extremity of the Moss Landing Harbor North Breakwater.

80.1225 Santa Cruz Harbor, CA. A line drawn from the seaward extremity of the Santa Cruz Harbor East Breakwater to Santa Cruz Harbor West Breakwater Light; thence to Santa Cruz Light.

80.1230 Pillar Point Harbor, CA. A line drawn from Pillar Point Harbor Light 6 to Pillar Point Harbor Entrance Light.

80.1250 San Francisco Harbor, CA. A straight line drawn from Point Bonita Light through Mile Rocks Light to the shore.

80.1255 Bodega and Tomales Bay, CA. (a) An east-west line drawn from Sand Point to Avalis Beach.

(b) A line drawn from the seaward extremity of Bodega Harbor North Breakwater to Bodega Harbor Entrance Light 1. [CGD 81-017, 46 FR 28154, May 26, 1981]

80.1260 Albion River, CA. A line drawn on an axis of 030° true through Albion River Light 1 across Albion Cove.

80.1265 Noyo River, CA. A line drawn from Noyo River Entrance Daybeacon 4 to Noyo River Entrance Light 5.

80.1270 Arcata-Humboldt Bay, CA. A line drawn from Humboldt Bay Entrance Light 4 to Humboldt Bay Entrance Light 3.

80.1275 Crescent City Harbor, CA. A line drawn from Crescent City Entrance Light to the southeasternmost extremity of Whaler Island.

Thirteenth District

80.1305 Chetco River, OR. A line drawn across the seaward extremities of the Chetco River Entrance Jetties. [CGD 81-017, 46 FR 28154, May 26, 1981]

80.1310 Rogue River, OR. A line drawn across the seaward extremities of the Rogue River Entrance Jetties.

80.1315 Coquille River, OR. A line drawn across the seaward extremities of the Coquille River Entrance Jetties.

80.1320 Coos Bay, OR. A line drawn across the seaward extremities of the Coos Bay Entrance Jetties.

80.1325 Umpqua River, OR. A line drawn across the seaward extremities of the Umpqua Entrance Jetties.

80.1330 Siuslaw River, OR. A line drawn across the seaward extremities of the Siuslaw River Entrance Jetties.

80.1335 Alsea Bay, OR. A line drawn from the seaward shoreline on the north of the Alsea Bay Entrance 165° true across the channel entrance.

80.1340 Yaquina Bay, OR. A line drawn across the seaward extremities of Yaquina Bay Entrance Jetties.

80.1345 Depoe Bay, OR. A line drawn across the Depoe Bay Channel entrance parallel with the general trend of the highwater shoreline.

80.1350 Netarts Bay, OR. A line drawn from the northernmost extremity of the shore on the south side of Netarts Bay north to the opposite shoreline.

80.1355 Tillamook Bay, OR. A line drawn across the seaward extremities of the Tillamook Bay Entrance Jetties.

80.1360 Nehalem River, OR. A line drawn approximately parallel with the general trend of the highwater shoreline across the Nehalem River Entrance.

80.1365 Columbia River Entrance, OR/WA. A line drawn from the seaward extremity of the Columbia River North Jetty (above water) 155° true to the seaward extremity of the Columbia River South Jetty (above water).

80.1370 Willapa Bay, WA. A line drawn from Willapa Bay Light 169.8° true to the westernmost tripod charted 1.6 miles south of Leadbetter Point.

80.1375 Grays Harbor, WA. A line drawn across the seaward extremities (above water) of the Grays Harbor Entrance Jetties.

80.1380 Quillayute River, WA. A line drawn from the seaward extremity of the Quillayute River Entrance East Jetty to the overhead power cable tower charted on James Island; thence a straight line through Quillayute River Entrance Light 3 to the shoreline.

80.1385 Strait of Juan de Fuca. The 72 COLREGS shall apply on all waters of the Strait of Juan de Fuca.

80.1390 Haro Strait and Strait of Georgia. The 72 COLREGS shall apply on all waters of the Haro Strait and the Strait of Georgia.

80.1395 Puget Sound and Adjacent Waters. The 72 COLREGS shall apply on all waters of Puget Sound and adjacent waters, including Lake Union, Lake Washington, Hood Canal, and all tributaries. [CGD 81-087, 46 FR 61456, Dec. 17, 1981]

PACIFIC ISLANDS

Fourteenth District

80.1410 Hawaiian Island Exemption from General Rule. Except as provided elsewhere in this part for Mamala Bay and Kaneohe Bay on Oahu; Port Allen and Nawiliwili Bay on Kauai;

Kahului Harbor on Maui; and Kawailae and Hilo Harbors on Hawaii, the 72 COLREGS shall apply on all other bays, harbors, and lagoons of the Hawaiian Islands (including Midway).

80.1420 Mamala Bay, Oahu, HI. A line drawn from Barbers Point Light to Diamond Head Light.

80.1430 Kaneohe Bay, Oahu, HI. A straight line drawn from Pyramid Rock Light across Kaneohe Bay through the center of Mokolii Island to the shoreline.

80.1440 Port Allen, Kauai, HI. A line drawn from Hanapepe Light to Hanapepe Bay Breakwater Light. [CGD 81-017, 46 FR 28154, May 26, 1981]

80.1450 Nawiliwili Harbor, Kauai, HI. A line drawn from Nawiliwili Harbor Breakwater Light to Kukii Point Light.

80.1460 Kahului Harbor, Maui, HI. A line drawn from Kahului Harbor Entrance East Breakwater Light to Kahului Harbor Entrance West Breakwater Light.

80.1470 Kawaihae Harbor, Hawaii, HI. A line drawn from Kawaihae Light to the seaward extremity of the Kawaihae South Breakwater.

80.1480 Hilo Harbor, Hawaii, HI. A line drawn from the seaward extremity of the Hilo Breakwater 265° true (as an extension of the seaward side of the breakwater) to the shoreline 0.2 nautical mile north of Alealea Point.

80.1490 Apra Harbor, U.S. Territory of Guam. A line drawn from the westernmost extremity of Orote Island to the westernmost extremity of Glass Breakwater.

80.1495 U.S. Pacific Island Possessions. The 72 COLREGS shall apply on the bays, harbors, lagoons, and waters surrounding the U.S. Pacific Island Possessions of American Samoa, Baker, Canton, Howland, Jarvis, Johnson, Palmyra, Swains and Wake Island. (The Trust Territory of the Pacific Islands is not a U.S. possession, and therefore Part 82 does not apply thereto.) [CGD 81-017, 46 FR 28154, May 26, 1981]

ALASKA

Seventeenth District

80.1705 Alaska. The 72 COLREGS shall apply on all the sounds, bays, harbors, and inlets of Alaska. [CGD 81-017, 46 FR 28154, May 26, 1981]

PENALTY PROVISIONS

VIOLATIONS OF INTERNATIONAL NAVIGATION RULES AND REGULATIONS (33 U.S.C. 1608)

Section 9.

(a) Whoever operates a vessel, subject to the provisions of this Act, in violation of this Act or of any regulation promulgated pursuant to section 8, shall be liable to a civil penalty of not more than $5,000 for each such violation.

(b) Every vessel subject to the provisions of this Act, other than a public vessel being used for noncommercial purposes, which is operated in violation of this Act or of any regulation promulgated pursuant to section 8, shall be liable to a civil penalty of not more than $5,000 for each such violation, for which penalty the vessel may be seized and proceeded against in the district court of the United States of any district within which such vessel may be found.

(c) The Secretary of the department in which the Coast Guard is operating may assess any civil penalty authorized by this section. No such penalty maybe assessed until the person charged, or the owner of the vessel charged, as appropriate, shall have been given notice of the violation involved and an opportunity for a hearing. For good cause shown, the Secretary may remit, mitigate, or compromise any penalty assessed. Upon the failure of the person charged, or the owner of the vessel charged, to pay an assessed penalty, as it may have been mitigated or compromised, the Secretary may request the Attorney General to commence an action in the appropriate district court of the United States for collection of the penalty as assessed, without regard to the amount involved, together with such other relief as may be appropriate. (July 27, 1977, Section 9, 91 Stat. 310)

VIOLATIONS OF INLAND NAVIGATION RULES AND REGULATIONS (33 U.S.C. 2072)

Section 4.

(a) Whoever operates a vessel in violation of this Act, or of any regulation issued thereunder, or in violation of a certificate of alternative compliance issued under Rule 1 is liable to a civil penalty of not more than $5,000 for each violation.

(b) Every vessel subject to this Act, other than a public vessel being used for noncommercial purposes, that is operated in violation of this Act, or of any regulation issued thereunder, or in violation of a certificate of alternative compliance issued under Rule 1 is liable to a civil penalty of not more than $5,000 for each violation, for which penalty the vessel may be seized and proceeded against in the district court of the United States of any district within which the vessel may be found.

(c) The Secretary may assess any civil penalty authorized by this section. No such penalty may be assessed until the person charged, or the owner of the vessel charged, as appropriate, shall have been given notice of the violation involved and an opportunity for a hearing. For good cause shown, the Secretary may remit, mitigate, or compromise any penalty assessed. Upon the failure of the person charged, or the owner of the vessel charged, to pay an assessed penalty, as it may have been mitigated or compromised, the Secretary may request the Attorney General to commence an action in the appropriate district court of the United States for collection of the penalty as assessed, without regard to the amount involved, together with such other relief as may be appropriate.

(d) The Secretary of the Treasury shall withhold or revoke, at the request of the Secretary, the clearance, required by section 4197 of the Revised Statutes of the United States (46 U.S.C. 91) of any vessel, the owner or operator of which is subject to any of the penalties in this section. Clearance may be granted in such cases upon the filing of a bond or other surety satisfactory to the Secretary. (Dec. 24, 1980, Section 4, 94 Stat. 3433)

OTHER PENALTIES AND DUTIES
specified in 46 U.S.C. 2301-2305

Excerpt from Title 46 of the United States Code
CHAPTER 23 - OPERATIONS OF VESSELS GENERALLY
[Enacted on August 26, 1983]

2301 Application.
2302 Penalties for negligent operations.
2303 Duties related to marine casualty assistance.
2304 Duty to provide assistance at sea.
2305 Injunctions.

2301 Application.
This chapter applies to a vessel operated on waters subject to the jurisdiction of the United States and, for a vessel owned in the United States, on the high seas.

2302 Penalties for negligent operations.
(a) A person operating a vessel in a negligent manner that endangers the life, limb, or property of a person is liable to the United States Government for a civil penalty of not more than $1,000.
(b) A person operating a vessel in a grossly negligent manner that endangers the life, limb, or property of a person shall be fined not more than $5,000, imprisoned for not more than one year, or both.
(c) For a penalty imposed under this section, the vessel also is liable in rem unless

the vessel is
  (1) owned by a State or a political subdivision of a State;
  (2) operated principally for government purposes; and
  (3) identified clearly as a vessel of that State or subdivision

2303 Duties related to marine casualty assistance and information

(a) The master or individual in charge of a vessel involved in a marine casualty shall:

  (1) render necessary assistance to each individual affected to save that affected individual from danger caused by the marine casualty, so far as the master or individual in charge can do so without serious danger to the master's or individual's vessel or to individuals on board; and

  (2) give the master's or individual's name and address and identification of the vessel to the master or individual in charge of any other vessel involved in the casualty, to any individual injured, and to the owner of any property damaged.

(b) An individual violating this section or a regulation prescribed under this section shall be fined not more than $1,000 or imprisoned for not more than 2 years. The vessel also is liable in rem to the United States Government for the fine.

(c) An individual complying with subsection (a) of this section or gratuitously and in good faith rendering assistance at the scene of a marine casualty without objection by an individual assisted, is not liable for damages as a result of rendering assistance or for an act or omission in providing or arranging salvage, towage, medical treatment, or other assistance when the individual acts as an ordinary, reasonable, and prudent individual would have acted under the circumstances.

2304 Duty to provide assistance at sea.

(a) A master or individual in charge of a vessel shall render assistance to any individual found at sea in danger of being lost, so far as the master or individual in charge can do so without serious danger to the master's or individual's vessel or individuals on board.

(b) A master or individual violating this section shall be fined not more than $1,000, imprisoned for not more than 2 years, or both.

2305 Injunctions.

(a) The district courts of the United States have jurisdiction to enjoin the negligent operation of vessels prohibited by this chapter on the petition of the Attorney General for the United States Government.

(b) When practicable, the Secretary shall:

  (1) give notice to any person against whom an action for injunctive relief is considered under this section an opportunity to present that person's views; and

  (2) except for a knowing and willful violation, give the person a reasonable opportunity to achieve compliance.

(c) The failure to give notice and opportunity to present views under subsection (b) of this section does not preclude the court from granting appropriate relief.

ALTERNATIVE COMPLIANCE

The alternative compliance procedures for the International Rules and the Inland Rules are the same, although they appear both in the International Rules section of the Code of Federal Regulations (33 CFR Part 81) and in the Inland Rules section (33 CFR Part 89)

1. Definitions.
2. General.
3. Application for a certificate of alternative compliance.
4. Certificate of alternative compliance: Contents
5. Certificate of alternative compliance: Termination.
6. Record of certification of vessels of special construction or purpose.

1. Definitions.

As used in this part:

"72 COLREGS" refers to the International Regulations for Preventing Collisions at Sea, 1972, done at London, October 20, 1972, as rectified by the Proces-Verbal of December 1, 1973 amended.

"Inland Rules" refers to the Inland Navigation Rules contained in the Inland Navigational Rules Act of 1980 (Pub. L 96-591) and the technical annexes established under that act.

"A vessel of special construction or purpose" means a vessel designed or modified to perform a special function and whose arrangement is thereby made relatively inflexible.

"Interference with the special function of the vessel" occurs when installation or use of lights, shapes, or sound-signaling appliances under the 72 COLREGS/ Inland Rules prevents or significantly hinders the operation in which the vessel is usually engaged.

2. General.

Vessels of special construction or purpose which cannot fully comply with the light, shape, and sound signal provisions of the 72 COLREGS/Inland Rules without interfering with their special function may instead meet alternative requirements. The Chief of the Marine Safety Division in each Coast Guard District Office makes this determination and requires that alternative compliance be as close as possible with the 72 COLREGS/Inland Rules. These regulations set out the procedure by which a vessel may be certified for alternative compliance.

3. Application for a Certificate of Alternative Compliance.

(a) The owner, builder, operator, or agent of a vessel of special construction or purpose who believes the vessel cannot fully comply with the 72 COLREGS/Inland Rules light, shape, or sound signal provisions without interference with its special function may apply for a determination that alternative compliance is justified. The application must be in writing, submitted to the Chief of the Marine Safety Division of the Coast Guard District in which the vessel is being built or operated, and include the following information:

(1) The name, address, and telephone number of the applicant.

(2) The identification of the vessel by its:

(i) Official number;

(ii) Shipyard hull number;

(iii) Hull identification number; or

(iv) State number, if the vessel does not have an official number or hull identification number.

(3) Vessel name and home port, if known.

(4) A description of the vessel's area of operation.

(5) A description of the provision for which the Certificate of Alternative Compliance is sought, including:

(i) The 72 COLREGS/Inland Rules Rule or Annex section number for which the Certificate of Alternative Compliance is sought;

(ii) A description of the special function of the vessel that would be interfered with by full compliance with the provision of that Rule or Annex section; and

(iii) A statement of how full compliance would interfere with the special function of the vessel.

(6) A description of the alternative installation that is in closest possible compliance with the applicable 72 COLREGS/Inland Rules Rule or Annex section.

(7) A copy of the vessel's plans or an accurate scale drawing that clearly shows:

(i) The required installation of the equipment under the 72 COLREGS / Inland Rules,

(ii) The proposed installation of the equipment for which certification is being sought, and

(iii) Any obstructions that may interfere with the equipment when installed in: (A) the required location; and (B) the proposed location.

(b) The Coast Guard may request from the applicant additional information concerning the application.

4. Certificate of Alternative Compliance: Contents.

The Chief of the Marine Safety Division issues the Certificate of Alternative Compliance to the vessel based on a determination that it cannot comply fully with 72 COLREGS/Inland Rules light, shape, and sound signal provisions without interference with its special function. This Certificate includes:

(a) Identification of the vessel as supplied in the application;

(b) The provision of the 72 COLREGS/Inland Rules for which the Certificate authorizes alternative compliance;

(c) A certification that the vessel is unable to comply fully with the 72 COLREGS/ Inland Rules light, shape, and sound signal requirements without interference with its special function;

(d) A statement of why full compliance would interfere with the special function of the vessel;

(e) The required alternative installation;

(f) A statement that the required alternative installation is in the closest possible compliance with the 72 COLREGS/Inland Rules without interfering with the special function of the vessel;

(g) The date of issuance;

(h) A statement that the Certificate of Alternative Compliance terminates when the vessel ceases to be usually engaged in the operation for which the certificate is issued.

5. Certificate of Alternative Compliance: Termination. The Certificate of Alternative Compliance terminates if the information supplied under 3.(a) or the Certificate issued under 4. is no longer applicable to the vessel.

6. Record of certification of vessels of special construction or purpose.

(a) Copies of Certificates of Alternative Compliance and documentation concerning Coast Guard Vessels are available for inspection at Coast Guard Headquarters, Office of Navigation Safety and Waterway Services, 2100 Second Street, S.W., Washington, D.C. 20593.

(b) The owner or operator of a vessel issued a Certificate shall ensure that the vessel does not operate unless the Certificate of Alternative Compliance or a certified copy of that Certificate is on board the vessel and available for inspection by Coast Guard personnel.

WATERS SPECIFIED BY THE SECRETARY
Waters Upon Which Certain Inland Navigation Rules Apply (33 CFR)

33 CFR 89.25
Waters Upon Which Inland Rules 9(a)(ii), 14(d), and 15(b) apply.
Inland Rules 9(a)(ii), 14(d), and 15(b) apply on the Great Lakes, the Western Rivers, and the following specified waters:

    (a) Tennessee-Tombigbee Waterway;

    (b) Tombigbee River;

    (c) Black Warrior River;

    (d) Alabama River;

    (e) Coosa River;

    (f) Mobile River above the Cochrane Bridge at St. Louis Point;

(g) Flint River;

(h) Chattachoochee River; and

(i) The Apalachicola River above its confluence with the Jackson River.

33 CFR §89.27
Waters upon which Inland Rule 24(i) applies.

(a) Inland Rule 24(i) applies on the Western Rivers and the specified waters listed in §89.25 (a) through (i).

(b) Inland Rule 24(i) applies on the Gulf Intracoastal Waterway from St. Marks, Florida, to the Rio Grande, Texas, including the Morgan City-Port Allen Alternate Route and the Galveston-Freeport Cutoff, except that a power-driven vessel pushing ahead or towing alongside shall exhibit the lights required by Inland Rule 24(c), while transiting within the following areas:

(1) St. Andrews Bay from the Hathaway Fixed Bridge at Mile 284.6 East of Harvey Locks (EHL) to the DuPont Fixed Bridge at Mile 295.4 EHL.

(2) Pensacola Bay, Santa Rosa Sound and Big Lagoon from the Light "10" off of Trout Point at Mile 176.9 EHL to the Pensacola Fixed Bridge at Mile 189.1 EHL.

(3) Mobile Bay and Bon Secour Bay from the Dauphin Island Causeway Fixed Bridge at Mile 127.7 EHL to Little Point Clear at Mile 140 EHL.

(4) Mississippi Sound from Grand Island Waterway Light "1" at Mile 53.8 EHL to Light "40" off the West Point of Dauphin Island at Mile 118.7 EHL.

(5) The Mississippi River at New Orleans, Mississippi River Gulf-Outlet Canal and the Inner Harbor Navigation Canal from the junction of the Harvey Canal and the Algiers Alternate Route at Mile 6.5 West of Harvey Locks (WHL) to the Michoud Canal at Mile 18 EHL.

(6) The Calcasieu River from the Calcasieu Lock at Mile 238.6 WHL to the Ellender Lift Bridge at Mile 243.6 WHL.

(7) The Sabine Neches Canal from Mile 262.5 WHL to Mile 291.5 WHL.

(8) Bolivar Roads from the Bolivar Assembling Basin at Mile 346 WHL to the Galveston Causeway Bridge at Mile 357.3 WHL.

(9) Freeport Harbor from Surfside Beach Fixed Bridge at Mile 393.8 WHL to the Bryan Beach Pontoon Bridge at Mile 397.6 WHL.

(10) Matagorda Ship Channel area of Matagorda Bay from Range "K" Front Light at Mile 468.7 WHL to the Port O'Connor Jetty at Mile 472.2 WHL.

(11) Corpus Christi Bay from Redfish Bay Day Beacon "55" at Mile 537.4 WHL when in the Gulf Intracoastal Waterway main route or from the north end of Lydia Ann Island Mile 531.1A when in the Gulf Intracoastal Waterway Alternate Route to Corpus Christi Bay LT 76 at Mile 543.7 WHL.

(12) Port Isabel and Brownsville Ship Channel south of the Padre Island Causeway Fixed Bridge Mile 665.1 WHL.

VESSEL BRIDGE-TO-BRIDGE RADIOTELEPHONE REGULATIONS

The Vessel Bridge-to-Bridge Radiotelephone Act is applicable on navigable waters of the United States inside of the boundary lines established in 46 CFR 7. In all cases, the Act applies on waters subject to the Inland Rules. In some instances, the Act may apply all the way out to the three miles limit, depending on where the boundary lines are located. In no instance does the Act apply beyond the three mile limit.

contents

26.01 Purpose.
26.02 Definitions.
26.03 Radiotelephone required.
26.04 Use of the designated frequency.
26.05 Use of radiotelephone.
26.06 Maintenance of radiotelephone; failure of radiotelephone.
26.07 Communications.
26.08 Exemption procedures.
26.09 List of exemptions.
26.10 Penalties.

AUTHORITY:33 U.S.C. 1201 - 1208; 49 CFR 1.46. §26.04 and §26.90 also issued under Sec. 4118, Pub. L. 101-380, 104 Stat. 523 (33 U.S.C. §1203 note).

SOURCE: CGD 71-114R, 37 FR 12720, June 28, 1972, unless otherwise noted.

26.01 Purpose.

(a) The purpose of this part is to implement the provisions of the Vessel Bridge-to-Bridge Radiotelephone Act. This part:
   (1) Requires the use of the vessel bridge-to-bridge radiotelephone;
   (2) Provides the Coast Guard's interpretation of the meaning of important terms in the Act;
   (3) Prescribes the procedures for applying for an exemption from the Act and the regulations issued under the Act and a listing of exemptions.
(b) Nothing in this part relieves any person from the obligation of complying with the rules of the road and the applicable pilot rules.

26.02 Definitions.

For the purpose of this part and interpreting the Act:
"Secretary" means the Secretary of the Department in which the Coast Guard is operating;
"Act" means the "Vessel Bridge-to-Bridge Radiotelephone Act", 33 U.S.C. Sections 1201-1208;
"Length" is measured from end to end over the deck excluding sheer;
"Power-driven vessel" means any vessel propelled by machinery; and

"Towing vessel" means any commercial vessel engaged in towing another vessel astern, alongside, or by pushing ahead.

"Vessel Traffic Services" (VTS) means a service implemented under part 161 of this chapter by the United States Coast Guard designed to improve the safety and efficiency of vessel traffic and to protect the environment. The VTS has the capability to interact with marine traffic and respond to traffic situations developing in the VTS area.

"Vessel Traffic Service Area or VTS Area" means the geographical area encompassing a specific VTS area of service as described in Part 161 of this chapter. This area of service may be subdivided into sectors for the purpose of allocating responsibility to individual Vessel Traffic Centers or to identify different operating requirements.

Note: Although regulatory jurisdiction is limited to the navigable waters of the United States, certain vessels will be encouraged or may be required, as a condition of port entry, to report beyond this area to facilitate traffic management within the VTS area.

(Rule 1, International Regulations for Preventing collisions at sea, 1972 (as rectified); EO 11964 (14 U.S.C. 2); 49 CFR 1.46(b)) [CGD 71-114R, 37 FR 12720, June 28, 1972, as amended by CGD 77-118a, 42 FR 35784, July 11, 1977; CGD 90-020, 59 FR 36313, July 15, 1994]

26.03 Radiotelephone required.

(a) Unless an exemption is granted under 26.09 and except as provided in paragraph (a)(4) of this section, this part applies to:

(1) Every power-driven vessel of 20 meters or over in length while navigating;

(2) Every vessel of 100 gross tons and upward carrying one or more passengers for hire while navigating;

(3) Every towing vessel of 26 feet or over in length while navigating; and

(4) Every dredge and floating plant engaged in or near a channel or fairway in operations likely to restrict or affect navigation of other vessels except for an unmanned or intermittently manned floating plant under the control of a dredge.

(b) Every vessel, dredge, or floating plant described in paragraph (a) of this section must have a radiotelephone on board capable of operation from its navigational bridge, or in the case of a dredge, from its main control station, and capable of transmitting and receiving on the frequency or frequencies within the 156-162 Mega-Hertz band using the classes of emissions designated by the Federal Communications Commission for the exchange of navigational information.

(c) The radiotelephone required by paragraph (b) of this section must be carried on board the described vessels, dredges, and floating plants upon the navigable waters of the United States.

(d) The radiotelephone required by paragraph (b) of this section must be capable of transmitting and receiving on VHF FM channel 22A (157.1 MHz).

(e) While transiting any of the following waters, each vessel described in paragraph (a) of this section also must have on board a radiotelephone capable of transmitting and receiving on VHF FM channel 67 (156.375 MHz):

(1) The lower Mississippi River from the territorial sea boundary, and within either the Southwest Pass safety fairway or the South Pass safety fairway specified in 33 CFR 166.200, to mile 242.4 AHP (Above Head of Passes) near Baton Rouge;

(2) The Mississippi River-Gulf Outlet from the territorial sea boundary, and within the Mississippi River-Gulf Outlet safety fairway specified in 33 CFR 166.200, to that channel's junction with the Inner Harbor Navigation Canal; and

(3) The full length of the Inner Harbor Navigation Canal from its junction with the Missisippi River to that canal's entry to Lake Pontchartrain at the New Seabrook vehicular bridge. [CGD 71-114R, 37 FR 12720, June 28, 1972, as amended by CGD 91-046, 47 FR 14483, Apr. 21, 1992]

(f) In addition to the radiotelephone required by paragraph (b) of this section, each vessel described in paragraph (a) of this section while transiting any waters within a Vessel Traffic Service Area, must have on board a radiotelephone capable of transmitting and receiving on the VTS designated frequency in Table 26.03(f) (VTS Call Signs, Designated Frequencies, and Monitoring Areas), located just following this section.

Note: A single VHF-FM radio capable of scanning or sequential monitoring (often referred to as "dual watch" capability) will not meet the requirements for two radios. And add to §26.03 Citations: CGD 90-020, 59 FR 36316, July 15, 1994

26.04 Use of the designated frequency.

(a) No person may use the frequency designated by the Federal Communications Commission under section 8 of the Act, 33 U.S.C. 1207(a), to transmit any information other than information necessary for the safe navigation of vessels or necessary tests.

(b) Each person who is required to maintain a listening watch under section 5 of the Act shall, when necessary, transmit and confirm, on the designated frequency, the intentions of his vessel and any other information necessary for the safe navigation of vessels.

(c) Nothing in these regulations may be construed as prohibiting the use of the designated frequency to communicate with shore stations to obtain or furnish information necessary for the safe navigation of vessels.

(d) On the navigable waters of the United States, channel 13 (156.65 MHz) is the designated frequency required to be monitored in accordance with section 26.05(a) except that in the area prescribed in section 26.03(e), channel 67 (156.375 MHz) is the designated frequency.

(e) On those navigable waters of the United States within a VTS area, the designated VTS frequency is the designated frequency required to be monitored in accordance with §26.05.

Note: As stated in 47 CFR 80.148(b), a VHF watch on Channel 16 (156.800MHz) is not required on vessels subject to the Vessel Bridge-to-Bridge Radiotelephone Act and participating in a Vessel Traffic Service (VTS) system when the watch is maintained on both the vessel bridge-to-bridge frequency and a designated VTS frequency.

[CGD 71-114R, 37 FR 12720, June 28, 1972, as ammended by CGD 91-046, 47 FR 14483, Apr. 21, 1992: CGD 90-020, 59 FR 36316, July 15, 1994]

26.05 Use of radiotelephone.
Section 5 of the Act states: (a) The radiotelephone required by this Act is for the exclusive use of the master or person in charge of the vessel, or the person designated by the master or person in charge to pilot or direct the movement of the vessel, who shall maintain a listening watch on the designated frequency. Nothing contained herein shall be interpreted as precluding the use of portable radiotelephone equipment to satisfy the requirements of this Act.

26.06 Maintenance of radiotelephone, failure of radiotelephone.
Section 6 of the Act states: (a) Whenever radiotelephone capability is required by this Act, a vessel's radiotelephone equipment shall be maintained in effective operating condition. If the radiotelephone equipment carried aboard a vessel ceases to operate, the master shall exercise due diligence to restore it or cause it to be restored to effective operating condition at the earliest practicable time. The failure of a vessel's radiotelephone equipment shall not, in itself, constitute a violation of this Act, nor shall it obligate the master of any vessel to moor or anchor his vessel; however, the loss of radiotelephone capability shall be given consideration in the navigation of the vessel.

26.07 Communications.
No person may use the services of, and no person may serve as, a person required to maintain a listening watch under section 5 of the Act, 33 U.S.C. 1204, unless the person can communicate in the English language. [CGD 90-020, 59 FR 36316, July 15, 1994]

26.08 Exemption procedures.
(a) Any person may petition for an exemption from any provision of the Act or this part;
(b) Each petition must be submitted in writing to U.S. Coast Guard (G-NSR), 2100 Second Street, SW, Washington, D.C. 20593, and must state:
    (1) The provisions of the Act or this part from which an exemption is requested;

and

(2) The reasons why marine navigation will not be adversely affected if the exemption is granted and if the exemption relates to a local communication system how that system would fully comply with the intent of the concept of the Act but would not conform in detail if the exemption is granted.

[CGD 71-114R, 37 FR 12720, June 28, 1972, as amended by CGD 73-256, 39 FR 9176, Mar. 8, 1974]

26.09 List of exemptions.

(a) All vessels navigating on those waters governed by the navigation rules for the Great Lakes and their connecting and tributary waters (33 U.S.C. 241 et seq.) are exempt from the requirements of the Vessel Bridge-to-Bridge Radiotelephone Act and this part until May 6, 1975

(b) Each vessel navigating on the Great Lakes as defined in the Inland Navigational Rules Act of 1980 (33 U.S.C. 2001 et seq.) and to which the Vessel Bridge-to-Bridge Radiotelephone Act (33 U.S.C. 1201-1208) applies is exempt from the requirements in 33 U.S.C. 1203, 1204, and 1205 and the regulations under sections 26.03, 26.04, 26.05, 26.06, and 26.07. Each of these vessels and each person to whom 33 U.S.C. 1208(a) applies must comply with Articles VII, X, XI, XII, XIII, XV, and XVI and Technical Regulations 1-9 of "The Agreement Between the United States of America and Canada for Promotion of Safety on the Great Lakes by means of Radio, 1973." [CGD 72-23R, 37 FR 28633, Dec. 28, 1972, as amended by CGD 74-291, 39 FR 44980, Dec. 30, 1974; CGD 74-304, 40 FR 19470, May 5, 1975; CGD 83-003, 48 FR 7442, Feb. 22, 1983; CGD 91-046, 57 FR 14483, Apr. 21, 1992]

26.10 Penalties.

Section 9 of the Act states:

(a) Whoever, being the master or person in charge of a vessel subject to the Act, fails to enforce or comply with the Act or the regulations hereunder; or whoever, being designated by the master or person in charge of a vessel subject to the Act to pilot or direct the movement of a vessel fails to enforce or comply with the Act or the regulations hereunder is liable to a civil penalty of not more than $500 to be assessed by the Secretary.

(b) Every vessel navigated in violation of the Act or the regulations hereunder is liable to a civil penalty of not more than $500 to be assessed by the secretary, for which the vessel may be proceeded against in any District Court of the United States having jurisdiction.

(c) Any penalty assessed under this section may be remitted or mitigated by the Secretary, upon such terms as he may deem proper.

Table 26.03(F)
VESSEL TRAFFIC SERVICES (VTS) CALL SIGNS[1],
DESIGNATED FREQUENCIES[2],
AND MONITORING AREAS

NEW YORK

New York Traffic[3]   **156.700 MHz (Ch. 14)**

The waters of the Lower New York Bay west of a line drawn from Norton Point to Breezy Point and north of a line drawn from Ambrose Entrance Lighted Gong Buoy #1 to Ambrose Channel Lighted Gong Buoy #9 thence to West Bank Light and thence to Great Kills Light. The waters of the Upper New York Bay, south of 40°42.40'N. (Brooklyn Bridge) and 40°43.70'N. (Holland Tunnel Ventilator Shaft); and in Newark Bay, north of 40°38.25'N. (Arthur Kill Railroad Bridge), and south of 40°41.95'N. (Lehigh Valley Draw Bridge); and the Kill Van Kull.

New York Traffic[3]   **156.550 MHz (Ch. 11)**

The waters of Raritan Bay east of a line drawn from Great Kills Light to Point Comfort in New Jersey and south of a line drawn from Great Kills Light to West Bank Light to Ambrose Channel Lighted Gong Buoy #9 thence to Ambrose Channel Lighted Gong Buoy #1 and west of a line drawn from Ambrose Channel Lighted Gong Buoy #9, thence to Ambrose Channel Lighted Gong Buoy #1 and west of a line drawn from Ambrose Channel Lighted Gong Buoy #1 to the Sandy Hook Channel Entrance Buoys (Lighted Gong Buoys #1 and #2).

New York Traffic[3]   **156.600 MHz (Ch. 12)**

Each vessel at anchor within the above areas.

HOUSTON[3]

The navigable waters north of 29°N., west of 94°20'W., south of 29°49'N., and east of 95°20'W.:

Houston Traffic   **156.550 MHz (Ch. 11)**

The navigable waters north of a line extending due west from the southern most end of Exxon Dock #1 (29°43.37'N., 95°01.27'W.).

Houston Traffic   **156.600 MHz (Ch. 12)**

The navigable waters south of a line extending due west from the southern most end of Exxon Dock #1 (29°43.37'N., 95°01.27'W.).

BERWICK BAY

Berwick Traffic   156.550 MHz (Ch. 11)
The navigable waters south of 29°45'N., west of 91°10'W., north of 29°37'N., and east of 91°18'W.

ST. MARYS RIVER

Soo Control   156.600 MHz (Ch. 12)
The navigable waters of the St. Mary's River between 45°57'N. (De Tour Reef Light) and 46°38.7'N. (Ile Parisienne Light), except the St. Mary Falls Canal and those navigable waters east of a line from 46°04.16'N. and 46°01.57'N. (La Pointe to Sims Point in Potagannissing Bay and Worsley Bay).

SAN FRANCISCO[3]

San Francisco Traffic   156.600 MHz (Ch. 12)
The waters within a 38 nautical mile radius of Mount Tamalpais (37°55.8'N., 122°34.6'W.) excluding the San Francisco Offshore Precautionary Area.

San Francisco Traffic   156.700 MHz (Ch. 14)
The waters of the San Francisco Offshore Precautionary Area eastward to San Francisco Bay including its tributaries extending to the ports of Stockton, Sacramento and Redwood City.

PUGET SOUND[4]

Seattle Traffic[5]   156.700 MHz (Ch. 14)
The navigable waters of Puget Sound, Hood Canal and adjacent waters south of a line connecting Marrowstone Point and Lagoon Point in Admiralty Inlet and south of a line drawn due east from the southernmost tip of Possession Point on Whidbey Island to the shoreline.

Seattle Traffic[5]   156.250 MHz (Ch. 5A)
The navigable waters of the Strait of Juan de Fuca east of 124°40'W. Excluding the waters in the central portion of the Strait of Juan de Fuca north and east of Race Rocks; the navigable waters of the Strait of Georgia east of 122°52'W.; the San Juan Island Archipelago, Rosario Strait, Bellingham Bay; Admiralty Inlet north of a line connecting Marrowstone Point and Lagoon Point and all waters east of Whidbey Island north of a line drawn due east from the southernmost tip of Possession Point on Whidbey Island to the shoreline.

Tofino Traffic[6] **156.725 MHz (Ch. 74)**
The waters west of 124°40'W. Within 50 nautical miles of the coast of Vancouver Island including the waters north of 48°N and east of 127°W.

Vancouver Traffic **156.550 MHz (Ch. 11)**
The navigable waters of the Strait of Georgia west of 122°52'W., the navigable waters of the central Strait of Juan de Fuca north and east of Race Rocks, including the Gulf Island Archipelago, Boundary Pass and Haro Strait.

PRICE WILLIAM SOUND[7]

Valdez Traffic **156.650 MHz (Ch. 13)**
The navigable waters south of 61°05'N., east of 147°20'W., north of 60°N., and west of 146°30'W.; and, all navigable waters in Port Valdez.

LOUISVILLE[7]

Louisville Traffic **156.650 MHz (Ch. 13)**
The navigable waters of the Ohio River between McAlpine Locks (Mile 606) and Twelve Mile Island (Mile 593), only when the McAlpine upper pool gauge is at approximately 13.0 feet or above.

NOTES to Table 26.03 (f)

1. VTS regulations are denoted in 33 CFR Part 161. All geographic coordinates (latitude and longitude) are expressed in North American Datum of 1983 (NAD 83).

2. In the event of a communication failure either by the vessel traffic center or the vessel or radio congestion on a designated VTS frequency, communication may be established on an alternate VTS frequency. The bridge-to-bridge navigation frequency, 156.650 MHz (Channel 13), is monitored in each VTS area; and it may be used as an alternate frequency, however, only to the extent that doing so provides a level of safety beyond that provided by other means.

3. Designated frequency monitoring is required within U.S. navigable waters. In areas which are outside the U.S. navigable waters, designated frequency monitoring is voluntary. However, prospective VTS Users are encouraged to monitor the designated frequency.

4. A Cooperative Vessel Traffic Service was established by the United States and

Canada within adjoining waters. The appropriate vessel traffic center administers the rules issued by both nations; however, it will enforce only its own set of rules within its jurisdiction.

5. Seattle Traffic may direct a vessel to monitor the other primary VTS frequency 156.250 MHz or 156.700 MHz (Channel 5A or 14) depending on traffic density, weather conditions, or other safety factors, rather than strictly adhering to the designated frequency required for each monitoring area as defined above. This does not require a vessel to monitor both primary frequencies.

6. A portion of Tofino Sector's monitoring area extends beyond the defined CVTS area. Designated frequency monitoring is voluntary in these portions outside of VTS Jurisdiction, however, prospective VTS Users are encouraged to monitor the designated frequency.

7. The bridge-to-bridge navigational frequency, 156.650 MHz (Channel 13), is used in these VTSs because the level of radiotelephone transmissions does not warrant a designated VTS frequency. The listening watch required by §26.05 of this chapter is not limited to the monitoring area.

LEGAL CITATIONS

72 COLREGS

International Navigational Rules Act of 1977 .... Public Law 95-75;
91 Stat. 311; 33 U.S.C. 1601-1608
COLREGS Demarcation Lines .... 33 CFR 80
72 COLREGS, Implementing Rules .... 33 CFR 81
72 COLREGS, Interpretative Rules .... 33 CFR 82
Amendments to 72 COLREGS effective June 1, 1983 .... 48 FR 28634

Inland Rules

Inland Navigation Rules Act of 1980 .... Public Law 96-591;
94 Stat. 3415; 33 U.S.C. 2001-2038, 2071-2073.
Annex I, Positioning & Technical Details of Lights and Shapes ...... 33 CFR 84.
Annex II, Additional Signals for Fishing in Close Proximity .... 33 CFR 85.
Annex III, Technical Details of Sound Signal Appliances .... 33 CFR 86.
Annex IV, Distress Signals .... 33 CFR 87.
Annex V, Pilot Rules .... 33 CFR 88.
Inland Navigation Rules, Implementing Rules .... 33 CFR 89.
Inland Navigation Rules, Interpretative Rules .... 33 CFR 90.

Vessel Bridge-to-Bridge Radiotelephone
Vessel Bridge-to-Bridge Radiotelephone Act .... Public Law 92-63;
85 Stat. 164; 33 U.S.C. 1201-1208.
Vessel Bridge-to-Bridge Radiotelephone Regulations
(Coast Guard regulations) .... 33 CFR 26.
Radiotelephone Stations Provided for Compliance With the Vessel
Bridge-to-Bridge Radiotelephone Act
(Federal Communications Commission regulations) ... 47 CFR 83.701-83.725
Other FCC regulations pertaining to vessel bridge-to-bridge radiotelephone com-
munications are contained in various sections of 47 CFR 81 and 7 CFR 83.
Boundary Lines .... 46 CFR 7.

CONVERSION TABLE
Conversion of Metric to U.S. Customary/Imperial Units

Metric Measure	U.S. Customary/Imperial Measure
1000 Meters (M)	3280.8 ft.
500 M	1640.4 ft.
200 M	656.2 ft.
150 M	492.1 ft.
100 M	328.1 ft.
75 M	246.1 ft.
60 M	196.8 ft.
50 M	164.0 ft.
25 M	82.0 ft.
20 M	65.6 ft.
12 M	39.4 ft.
10 M	32.8 ft.
8 M	26.2 ft.
7 M	23.0 ft.
6 M	19.7 ft.
5 M	16.4 ft.
4.5 M	14.8 ft.
4.0 M	13.1 ft.
3.5 M	11.5 ft.
2.5 M	8.2 ft.
2.0 M	6.6 ft.
1.5 M	4.9 ft.
1 M	3.3 ft.
0.9 M	35.4 in.
0.6 M	23.6 in.
0.5 M	19.7 in.
300 Millimeters (mm)	11.8 in.
200 mm	7.9 in.

INTRODUCTION TO THE INLAND RULES

The Inland Rules in this book replace the old Inland Rules, Western Rivers Rules, Great Lakes Rules, their respective pilot rules and interpretive rules, and parts of the Motorboat Act of 1940. Many of the old navigation rules were originally inacted in the last century. Occasionally, provisions were added to cope with the increasing complexities of water transportation. Eventually the navigation rules for United States inland waterways became such a confusing patchwork of requirements that in the 1960's several attempts were made to revise and simplify them. These attempts were not successful.

Following the signing of the Convention on the International Regulations for Preventing Collisions at Sea, 1972, a new effort was made to unify and update the various inland navigation rules. This effort culminated in the enactment of the Inland Navigational Rules Act of 1980. This legislation sets out Rules 1 through 38—the main body of the Rules. The five Annexes were published as regulations. It is important to note that with the exception of Annex V to the Inland Rules, the International and Inland Rules and Annexes are very similar in both content and format.

The effective date for the Inland Navigation Rules was December 24, 1981, except for the Great Lakes where the effective date was March 1, 1983.

PART A — GENERAL

Rule 1. Application

(a) These Rules apply to all vessels upon the inland waters of the United States, and to vessels of the United States on the Canadian waters of the Great Lakes to the extent that there is no conflict with Canadian law.

(b) (i) These Rules constitute special rules made by an appropriate authority within the meaning of Rule 1(b) of the International Regulations.
(ii) All vessels complying with the construction and equipment requirements of the International Regulations are considered to be in compliance with these Rules.

(c) Nothing in these Rules shall interfere with the operation of any special rules made by the Secretary of the Navy with respect to additional station or signal lights and shapes or whistle signals for ships of war and vessels proceeding under convoy, or by the Secretary with respect to additional station or signal lights and shapes for fishing vessels engaged in fishing as a fleet. These additional station or signal lights and shapes or whistle signals shall, so far as possible, be such that they cannot be mistaken for any light, shape, or signal authorized elsewhere under these Rules. Notice of such special rules shall be published in the Federal Register and, after the effective date specified in such notice they shall have effect as if they were a part

Note: The Inland Rules are the same as the International Rules except for...

of these Rules[1]. (*same footnote on submarines as in International Rules*)

(d) Traffic separation schemes may be established for the purposes of these Rules. Vessel traffic service regulations may be in effect in certain areas.

(e) Whenever the Secretary determines that a vessel or class of vessels of special construction or purpose cannot comply fully with the provisions of any of these Rules with respect to the number, position, range or arc of visibility of lights or shapes, as well as to the disposition and characteristics of sound-signalling appliances, the vessel shall comply with such other provisions in regard to the number, position, range or arc of visibility of lights or shapes, as well as to the disposition and characteristics of sound-signalling appliances, as the Secretary shall have determined to be the closest possible compliance with these Rules. The Secretary may issue a certificate of alternative compliance for a vessel or class of vessels specifying the closest possible compliance with these Rules. The Secretary of the Navy shall make these determinations and issue certificates of alternative compliance for vessels of the Navy.

(f) The Secretary may accept a certificate of alternative compliance issued by a contracting party to the International Regulations if he determines that the alternative compliance standards of the contracting party are substantially the same as those of the United States.

Rule 3. General Definitions

The general definitions for Inland Rules are the same as for international waters with the exception that the international definition of a vessel "constrained by Draft" (International Rule 3, item h) does not apply on inland waters and the following inland terms are added to the inland list of definitions:

(l) "Western Rivers" means the Mississippi River, its tributaries, South Pass, and Southwest Pass, to the navigational demarcation lines dividing the high seas from harbors, rivers, and other inland waters of the United States, and the port Allen-Morgan City Alternate Route, and that part of the Atachafalaya River above its junction with the Port Allen-Morgan City Alternate Route including the Old River and the Red River;

(m) "Great Lakes" means the Great Lakes and their connecting and tributary waters including the Calumet River as far as the Thomas J. O'Brien Lock and Controlling Works (between mile 326 and 327), the Chicago River as far as the east side of the Ashland Avenue Bridge (between mile 321 and 322), and the Saint Lawrence River as far east as the lower exit of Saint Lambert Lock;

(n) "Secretary" means the Secretary of the department in which the Coast Guard is operating;

(o) "Inland Waters" means the navigable waters of the United States shoreward of the navigational demarcation lines dividing the high seas from harbors, rivers, and other inland waters of the United States and the waters of the Great Lakes on the United States side of the International Boundary;

(p) "Inland Rules" or "Rules" mean the Inland Navigational Rules and the annexes thereto, which govern the conduct of vessels and specify the lights, shapes and sound signals that apply on inland waters; and

(q) "International Regulations" means the International Regulations for Preventing Collisions at Sea, 1972, including annexes currently in force in the United States.

PART B — STEERING AND SAILING RULES
Subpart I — Conduct of Vessels in Any Condition of Visibility

Rule 9. Narrow Channels

(a) (i) *is the same as International Rule 9 Part (a).*

(a) (ii) Notwithstanding paragraph (a)(i) and Rule 14(a), a power-driven vessel operating in narrow channels or fairways on the Great Lakes, Western Rivers, or waters specified by the Secretary, and proceeding downbound with a following current shall have the right-of-way over an upbound vessel, shall propose the manner and place of passage, and shall initiate the maneuvering signals prescribed in Rule 34(a)(i), as appropriate. The vessel proceeding upbound against the current shall hold as necessary to permit safe passing. *Editor's note: see page 52 for waters specified by the Secretary.*

(d) A vessel shall not cross a narrow channel or fairway if such crossing impedes the passage of a vessel which can safely navigate only within that channel or fairway. The latter vessel shall use the danger signal prescribed in Rule 34(d) if in doubt as to the intention of the crossing vessel.

(e) (i) In a narrow channel or fairway when overtaking, the vessel intending to overtake shall indicate her intention by sounding the appropriate signal prescribed in Rule 34(c) and take steps to permit safe passing. The overtaken vessel, if in agreement, shall sound the same signal. If in doubt she shall sound the danger signal prescribed in Rule 34(d).

Rule 10. Traffic Separation Schemes

(a) This Rule applies to traffic separation schemes and does not relieve any vessel of her obligation under any other Rule.

Otherwise Rule 10 is the same as International Rule 10.

Note: The Inland Rules are the same as the International Rules except for...

PART B — STEERING AND SAILING RULES
Subpart II — Conduct of Vessels in sight of one another

RULE 14. Head-on Situation

(a) Unless otherwise agreed, when two power-driven vessels are meeting on reciprocal or nearly reciprocal courses so as to involve risk of collision each shall alter her course to starboard so that each shall pass on the port side of the other.

(d) Not withstanding paragraph (a) of this Rule, a power-driven vessel operating on the Great Lakes, Western Rivers, or waters specified by the Secretary, and proceeding downbound with a following current shall have the right-of-way over an upbound vessel, shall propose the manner of passage, and shall initiate the maneuvering signals prescribed by Rule 34(a)(i), as appropriate.
Editor's note: see page 52 for waters specified by the Secretary.

Rule 15. Crossing Situation

(a) *Same as International Rule 15.*
(b) Notwithstanding paragraph (a), on the Great Lakes, Western Rivers, or water specified by the Secretary, a vessel crossing a river shall keep out of the way of a power-driven vessel ascending or descending the river.
Editor's note: see page 52 for waters specified by the Secretary.

Rule 18. Responsibility Between Vessels

Parts (d) (i) and (ii) on vessels constrained by draft to not apply to Inland Waters.

PART C — LIGHTS AND SHAPES

Rule 21. Definitions

(a) "Masthead light" means a white light placed over the fore and aft centerline of the vessel showing an unbroken light over an arc of the horizon of 225 degrees and so fixed as to show the light from right ahead to 22.5 degrees abaft the beam on either side of the vessel, except that on a vessel of less than 12 meters in length the masthead light shall be placed as nearly as practicable to the fore and aft centerline of the vessel.

(b) "Sidelights" mean a green light on the starboard side and a red light on the port side each showing an unbroken light over an arc of the horizon of 112.5 degrees and so fixed as to show the light from right ahead to 22.5 degrees abaft the beam on its respective side. On a vessel of less than 20 meters in length the sidelights may be combined in one lantern carried on the fore and aft centerline of the vessel, except that on a vessel of less than 12 meters in length the sidelights when combined in one lantern shall be placed as nearly as practicable to the fore and aft centerline of the vessel.

(g) "Special flashing light" means a yellow light flashing at regular intervals at a frequency of 50 to 70 flashes per minute, placed as far forward and as nearly as practicable on the fore and aft centerline of the tow and showing an unbroken light over an arc of the horizon of not less than 180 degrees nor more than 225 degrees and so fixed as to show the light from right ahead to abeam and no more than 22.5 degrees abaft the beam on either side of the vessel.

Rule 22. Visibility of Lights

Same as International Rule, except Parts (a), (b), and (c) each include:
—a special flashing light, 2 miles.

Rule 23. Power-driven Vessels Underway

(a) (i) a masthead light forward; except that a vessel of less than 20 meters in length need not exhibit this light forward of amidships but shall exhibit it as far forward as is practicable.

(c) (ii) *Does not apply to Inland waters*
(iii) *Does not apply to Inland waters*

(d) A power-driven vessel when operating on the Great Lakes may carry an all-round white light in lieu of the second masthead light and sternlight prescribed in paragraph (a) of this Rule. The light shall be carried in the position of the second masthead light and be visible at the same minimum range.

RULE 24. Towing and Pushing

(a) A power-driven vessel when towing astern shall exhibit:
(i) instead of the light prescribed either in Rule 23(a)(i) or 23(a)(ii), two masthead lights in a vertical line. When the length of the tow, measuring from the stern of the towing vessel to the after end of the tow exceeds 200 meters, three such lights in a vertical line;
(ii) sidelights;
(iii) a sternlight;
(iv) a towing light in a vertical line above the sternlight; and
(v) when the length of the tow exceeds 200 meters, a diamond shape where it can best be seen.

(b) When a pushing vessel and a vessel being pushed ahead are rigidly connected in a composite unit they shall be regarded as a power-driven vessel and exhibit the lights prescribed in Rule 23.

(c) A power-driven vessel when pushing ahead or towing alongside, except as required by paragraphs (b) and (i) of this Rule, shall exhibit:
>(i) instead of the light prescribed either in Rule 23(a)(i) or 23(a)(ii), two masthead lights in a vertical line;
>(ii) sidelights; and
>(iii) two towing lights in a vertical line.

(d) A power-driven vessel to which paragraphs (a) or (c) of this Rule apply shall also comply with Rule 23(a)(i) and 23(a)(ii).

(e) A vessel or object, other than those referred to in paragraph (g) of this Rule being towed shall exhibit:
>(i) sidelights;
>(ii) a sternlight; and
>(iii) when the length of the tow exceeds 200 meters, a diamond shape where it can best be seen.

(f) Provided that any number of vessels being towed alongside or pushed in a group shall be lighted as one vessel;
>(i) a vessel being pushed ahead, not being part of a composite unit, shall exhibit at the forward end sidelights, and a special flashing light; and
>(ii) a vessel being towed alongside shall exhibit a sternlight and at the forward end, sidelights.

(g) An inconspicuous, partly submerged vessel or object being towed shall exhibit:
>(i) if it is less than 25 meters in breadth, one all-round white light at or near each end;
>(ii) if it is 25 meters of more in breadth, four all-round white lights to mark its length and breadth;
>(iii) if it exceeds 100 meters in length, additional all-round white lights between the lights prescribed in subparagraphs (i) and (ii) so that the distance between the lights shall not exceed 100 meters: *Provided*, That any vessels or objects being towed alongside each other shall be lighted as one vessel or object;
>(iv) a diamond shape at or near the aftermost extremity of the last vessel or object being towed; and
>(v) the towing vessel may direct a searchlight in the direction of the tow to indicate its presence to an approaching vessel.

(h) Where from any sufficient cause it is impracticable for a vessel or object being towed to exhibit the lights prescribed in paragraph (e) or (g) of this Rule, all possible measures shall be taken to light the vessel or object towed or at least to indicate the presence of the unlighted vessel or object.

(i) Notwithstanding paragraph (c), on the Western Rivers (except below the Huey P. Long Bridge on the Mississippi River) and on waters specified by the Secretary, a power-driven vessel when pushing ahead or towing alongside, except as paragraph (b) applies, shall exhibit:

    (i) sidelights; and

    (ii) two towing lights in a vertical line.

Editor's note: see page 52 for waters specified by the Secretary.

(j) Where from any sufficient cause it is impracticable for a vessel not normally engaged in towing operations to display the lights prescribed by paragraph (a), (c), or (i) of this Rule, such vessel shall not be required to exhibit those lights when engaged in towing another vessel in distress or otherwise in need of assistance. All possible measures shall be taken to indicate the nature of the relationship between the towing vessel and the vessel being assisted. The searchlight authorized by Rule 36 may be used to illuminate the tow.

Rule 25. Sailing Vessels Underway and Vessels Under Oars

(e) A vessel proceeding under sail when also being propelled by machinery shall exhibit forward where it can best be seen a conical shape, apex downward. A vessel of less than 12 meters in length is not required to exhibit this shape, but may do so.

Rule 27. Vessels Not Under Command
or Restricted in their Ability to Maneuver

(c) A vessel engaged in a towing operation which severely restricts the towing vessel and her tow in their ability to deviate from their course shall, in addition to the lights or shapes prescribed in sub-paragraphs (b)(i) and (ii) of this Rule, exhibit the lights or shape prescribed in Rule 24.

Rule 28. Vessels Constrained by their Draft
This rule does not apply to Inland waters

Rule 30.   Anchored Vessels and Vessels Aground

(d) A vessel aground shall exhibit the lights prescribed in paragraph (a) or (b) of this Rule and in addition, if practicable, where they can best be seen:

    (i) two all-round red lights in a vertical line;

    (ii) three balls in a vertical line

(g) A vessel of less than 20 meters in length, when at anchor in a special anchorage area designated by the Secretary, shall not be required to exhibit the anchor lights and shapes required by this Rule.

PART D — SOUND AND LIGHT SIGNALS

Rule 34.   Maneuvering and Warning Signals

(a) When power-driven vessels are in sight of one another and meeting or crossing at a distance within half a mile of each other, each vessel underway, when maneuvering as authorized or required by these Rules:

(i) shall indicate that maneuver by the following signals on her whistle: one short blast to mean "I intend to leave you on my port side"; two short blasts to mean "I intend to leave you on my starboard side"; and three short blasts to mean "I am operating astern propulsion".

(ii) upon hearing the one or two blast signal of the other shall, if in agreement, sound the same whistle signal and take the steps necessary to effect a safe passing. If however, from any cause, the vessel doubts the safety of the proposed maneuver, she shall sound the danger signal specified in paragraph (d) of this Rule and each vessel shall take appropriate precautionary action until a safe passing agreement is made.

(b) A vessel may supplement the whistle signals prescribed in paragraph (a) of this Rule by light signals:

(i) These signals shall have the following significance: one flash to mean "I intend to leave you on my port side"; two flashes to mean "I intend to leave you on my starboard side"; three flashes to mean "I am operating astern propulsion";

(ii) The duration of each signal shall be about 1 second; and

(iii) The light used for this signal shall, if fitted, be one all-round white or yellow light, visible at a minimum range of 2 miles, synchronized with the whistle, and shall comply with the provisions of Annex I to these Rules.

(c) When in sight of one another:

(i) a power-driven vessel intending to overtake another power-driven vessel shall indicate her intention by the following signals on her whistle: one short blast to mean "I intend to overtake you on your starboard side"; two short blasts to mean "I intend to overtake you on your port side"; and

(ii) the power-driven vessel about to be overtaken shall, if in agreement, sound a similar sound signal. If in doubt she shall sound the danger signal prescribed in paragraph (d).

(g) When a power-driven vessel is leaving a dock or berth, she shall sound one prolonged blast.

(h) A vessel that reaches agreement with another vessel in a meeting, crossing, or overtaking situation by using the radiotelephone as prescribed by the Bridge-to-Bridge Radiotelephone Act (85 Stat. 165; 33 U.S.C. 1207), is not obliged to sound the whistle signals prescribed by this Rule, but may do so. If agreement is not reached, then whistle signals shall be exchanged in a timely manner and shall prevail.

Rule 35. Sound Signals in Restricted Visibility

With the exception of Part (j) below and the absense of the constrained by draft category on inland waters mentioned in Part (c), this rule is the same as the Internatonal Rule, although worded differently.

(c) *Same as International Rule except "constrained by draft" does not apply.*

(j) The following vessels shall not be required to sound signals as prescribed in paragraph (f) of this Rule when anchored in a special anchorage area designated by the Secretary:
    (i) a vessel of less than 20 meters in length; and
    (ii) a barge, canal boat, scow or other nondescript craft.

Rule 36. Signals to Attract Attention

If necessary to attract the attention of another vessel, any vessel may make light or sound signals that cannot be mistaken for any signal authorized elsewhere in these Rules, or may direct the beam of her searchlight in the direction of the danger, in such a way as not to embarrass any vessel.

Rule 37. Distress Signals

The distress signals for inland waters are the same as those for International waters with the following additional signal: a high intensity white light flashing at regular intervals from 50 to 70 times per minute.

PART E — EXEMPTIONS

Rule 38. Exemptions

Any vessel or class of vessels, the keel of which is laid or which is at a corresponding stage of construction before the date of enactment of this Act, provided that she complies with the requirements of:

(a) The Act of June 7, 1897 (30 Stat. 96), as amended (33 U.S.C. 154-232) for vessels navigating the waters subject to that statute;

(b) Section 4233 of the Revised Statutes (33 U.S.C. 301-356) for vessels navigating the waters subject to that statute;

(c) The Act of February 8, 1895 (28 Stat. 645), as amended (33 U.S.C. 241-295) for vessels navigating the waters subject to that statute; or

(d) Sections 3, 4, and 5 of the Act of April 25, 1940 (54 Stat. 163), as amended (46 U.S.C. 526 b, c, and d) for motorboats navigating the waters subject to that statute; shall be exempted from compliance with the technical Annexes to these Rules as follows:

(i) the installation of lights with ranges prescribed in Rule 22, until 4 years after the effective date of these Rules, except that vessels of less than 20 meters in length are permanently exempt;

(ii) the installation of lights with color specifications as prescribed in Annex I to these Rules, until 4 years after the effective date of these Rules, except that vessels of less than 20 meters in length are permanently exempt;

(iii) the repositioning of lights as a result of conversion to metric units and rounding off measurement figures, are permanently exempt; and

(iv) the horizontal repositioning of masthead lights prescribed by Annex I to these Rules:

(1) on vessels of less than 150 meters in length, permanent exemption.

(2) on vessels of 150 meters of more in length, until 9 years after the effective date of these Rules.

(v) the restructuring or repositioning of all lights to meet the prescriptions of Annex I to these Rules, until 9 years after the effective date of these Rules;

(vi) power-driven vessels of 12 meters or more but less than 20 meters in length are permanently exempt from the provisions of Rule 23(a)(i) and 23(a)(iv) provided that, in place of these lights, the vessel exhibits a white light aft visible all round the horizon; and

(vii) the requirements for sound signal appliances prescribed in Annex III to these Rules, until 9 years after the effective date of these Rules.

INLAND RULES ANNEX I
Positioning and Technical Details of Lights and Shapes

84.01 Definitions
(a) The term "height above the hull" means height above the uppermost continuous deck. This height shall be measured from the position vertically beneath the location of the light.

(b) The term "practical cut-off" means, for vessels 20 meters or more in length, 12.5 percent of the minimum luminous intensity (Table 84.15(b)) corresponding to the greatest range of visibility for which the requirements of Annex I are met.

(c) The term "Rule" or "Rules" means the Inland Navigation Rules contained in Sec. 2 of the Inland Navigational Rules Act of 1980 (Pub. L. 96-591, 94 Stat. 3415, 33 U.S.C. 2001, December 24, 1980 as amended.

84.03 Vertical positioning and spacing of lights
(a) On a power-driven vessel of 20 meters or more in length the masthead lights shall be placed as follows:

(1) The forward masthead light, or if only one masthead light is carried, then that light, at a height above the hull of not less than 5 meters, and, if the breadth of the vessel exceeds 5 meters, then at a height above the hull not less than such breadth, so however that the light need not be placed at a greater height above the hull than 8 meters;
(2) When two masthead lights are carried the after one shall be at least 2 meters vertically higher than the forward one.

(b) The vertical separation of the masthead lights of power-driven vessels shall be such that in all normal conditions of trim the after light will be seen over and separate from the forward light at a distance of 1000 meters from the stem when viewed from water level.

(c) The masthead light of a power-driven vessel of 12 meters but less than 20 meters in length shall be placed at a height above the gunwale of not less than 2.5 meters.

(d) The masthead light, or the all-round light described in Rule 23(c) of a power-driven vessel of less than 12 meters in length shall be carried at least one meter higher than the sidelights.

(e) One of the two or three masthead lights prescribed for a power-driven vessel when engaged in towing or pushing another vessel shall be placed in the same position as either the forward masthead light or the after masthead light, provided that the lowest after masthead light shall be at least 2 meters vertically higher than the highest forward masthead light.

(f) (1) The masthead light or lights prescribed in Rule 23(a) shall be so placed as to be above and clear of all other lights and obstructions except as described in paragraph (f)(2) of this section.

(2) When it is impracticable to carry the all-round lights prescribed in Rule 27(b)(i) below the masthead lights, they may be carried above the after masthead light(s) or vertically in between the forward masthead light(s) and after masthead light(s), provided that in the latter case the requirement of section 84.05(d) shall be complied with.

(g) The sidelights of a power-driven vessel shall be placed at least one meter lower than the forward masthead light. They shall not be so low as to be interfered with by deck lights.

(h) [Reserved]

(i) When the Rules prescribe two or three lights to be carried in a vertical line, they shall be spaced as follows:

(1) On a vessel of 20 meters in length or more such lights shall be spaced not less than 1 meter apart, and the lowest of these lights shall, except where a towing light is required, be placed at a height of not less than 4 meters above the hull;
(2) On a vessel of less than 20 meters in lenght such lights shall be spaced not less than 1 meter apart and the lowest of these lights shall, except where a towing light is required, be placed at a height of not less than 2 meters above the gunwale;
(3) When three lights are carried they shall be equally spaced.

(j) *Same as International Rules Annex I, Part 2 (j)*

(k) *Same as International Rules Annex I, Part 2 (k)*

84.05 Horizontal positioning and spacing of lights
(a) Except as specified in paragraph (b) of this section, when two masthead lights are prescribed for a power-driven vessel, the horizontal distance between them shall not be less than one quarter of the length of the vessel but need not be more than 50 meters. The forward light shall be placed not more than one half of the length of the vessel from the stem.

(b) On power-driven vessels 50 meters but less than 60 meters in length operated on the Western Rivers, and those waters specified in section 89.25, the horizontal distance between masthead lights shall not be less than 10 meters.

(c) On a power-driven vessel of 20 meters or more in length the sidelights shall not be placed in front of the forward masthead lights. They shall be placed at or near the side of the vessel.

(d) When the lights prescribed in Rule 27(b)(i) are placed vertically between the forward masthead light(s) and the after masthead light(s) these all-round lights shall be placed at a horizontal distance of not less than 2 meters from the fore and aft centerline of the vessel in the athwartship direction.

84.07 Details of location of direction-indicating lights for fishing vessels, dredgers and vessels engaged in underwater operations
Same as International Rules Annex I, Part 4.

84.09 Screens
(a) The sidelights of vessels of 20 meters or more in length shall be fitted with matt black inboard screens and meet the requirements of section 84.17. On vessels of less than 20 meters in length, the sidelights, if necessary to meet requirements of section 84.17, shall be fitted with matt black inboard screens. With a combined lantern using a single vertical filament and a very narrow division between the green and red sections, external screens need not be fitted.

(b) On power-driven vessels less than 12 meters in length constructed after July 31, 1983, the masthead light, or the all-round light described in Rule 23 (c) shall be screened to prevent direct illumination of the vessel forward of the operator's position.

84.11 Shapes
Same as International Rules Annex I, Part 6, except that a cylinder is not defined on inland waters.

84.13 Color specification of lights
(a) The chromaticity of all navigation lights shall conform to the following standards, which lie within the boundaries of the area of the diagram specified for each color by the International Commission on Illumination (CIE), in the "Colors of Light Signals", which is incorporated by reference. It is Publication CIE No. 2.2 (TC-1.6), 1975, and is available from the Illumination Engineering Society, 345 East 47th Street, New York, NY 10017. It is also available for inspection at the Office of the Federal Register, Room 8401, 1100 L Street N.W., Washington, D.C. 20408. This incorporation by reference was ap- proved by the Director of the Federal Register.

(b) *Same as corresponding table in International Rules Annex I, Part 7.*

84.15 Intensity of Lights
Same as International Rules Annex I, Part 8,.

84.17 Horizontal sectors
(a) *Same as International Rules Annex I, Part 9 (a).*

(b) All-round lights shall be so located as not to be obscured by masts, topmasts or structures within angular sectors of more than 6 degrees, except anchor lights prescribed in Rule 30, which need not be placed at an impracticable height above the hull, and the all-round white light described in Rule 23(d), which may not be obscured at all.

84.19 Vertical sectors
(a) The vertical sectors of electric lights as fitted, with the exception of lights on sailing vessels underway and on unmanned barges, shall ensure that:
    (1) *Same as International Rules Annex I, Part 10 (a) (i).*
    (2) *Same as International Rules Annex I, Part 10 (a) (ii).*

(b) *Same as International Rules Annex I, Part 10 (b)*

(c) In the case of unmanned barges the minimum required intensity of electric lights as fitted shall be maintained on the horizontal.

(d) In the case of lights other than electric lights these specifications shall be met as closely as possible.

84.21 Intensity of non-electric lights. *Same as International Annex I, Part 11.*

84.23 Maneuvering light
Notwithstanding the provisions of section 84.03(f), the maneuvering light described in Rule 34(b) shall be placed approximately in the same fore and aft vertical plane as the masthead light or lights and, where practicable, at a minimum height of one-half meter vertically above the forward masthead light, provided that it shall be carried not less than one-half meter vertically above or below the after masthead light. On a vessel where only one masthead light is carried the maneuvering light, if fitted, shall be carried where it can best be seen, not less than one-half meter vertically apart from the masthead light.

84.25 Approval. [Reserved]

INLAND RULES ANNEX II
Additional Signals for Fishing vessels Fishing in Close Proximity

85.1 General
The lights mentioned herein shall, if exhibited in pursuance of Rule 26(d), be placed where they can best be seen. They shall be at least 0.9 meter apart but at a lower level than lights prescribed in Rule 26(b)(i) and (c)(i) contained in Inland

Navigational Rules Act of 1980. The lights shall be visible all around the horizon at a distance of at least 1 mile but at a lesser distance than the lights prescribed by these Rules for fishing vessels.

85.3 Signals for trawlers.
Same as International Rules Annex II, Part 2.

85.5 Signals for purse seiners.
Same as International Rules Annex II, Part 3.

INLAND RULES ANNEX III
Technical Details of Sound Signal Appliances

Subpart A — Whistles

86.01 Frequencies and range of audibility.
The fundamental frequency of the signal shall lie within the range 70-525 Hz. The range of audibility of the signal from a whistle shall be determined by those frequencies, which may include the fundamental and/or one or more higher frequencies, which lie within the frequency ranges and provide the sound pressure levels specified in section 86.05.

86.03 Limits of fundamental frequencies.
To ensure a wide variety of whistle characteristics, the fundamental frequency of a whistle shall be between the following limits:

(a) 70-200 Hz, for a vessel 200 meters or more in length;

(b) 130-350 Hz, for a vessel 75 meters but less than 200 meters in length;

(c) 250-525 Hz, for a vessel less than 75 meters in length.

86.05 Sound signal intensity and range of audibility
A whistle on a vessel shall provide, in the direction of the forward axis of the whistle and at a distance of 1 meter from it, a sound pressure level in at least 1/3 - octave band of not less than the appropriate figure given in Table 86.05 within the following frequency ranges (+/- 1 percent):

(a) 130-1200 Hz, for a vessel 75 meters or more in length;

(b) 250-1600 Hz, for a vessel 20 meters but less than 75 meters in length;

(c) 250-2100 Hz, for a vessel 12 meters but less than 20 meters in length.

Table 86.05

Length of vessel in meters	Fundamental frequency range (HZ)	for measured frequencies (HZ)	1/3-octave band level at 1 meter in dB referred to 2×10^{-5} N/m^2	Audibility range in nautical miles
200 or more	70-200	130-180	145	2
		180-250	143	
		250-1200	140	
75 but less than 200	130-250	130-180	140	1.5
		180-250	138	
		250-1200	134	
20 but less than 75	180-250	250-450	130	1.0
		450-800	125	
		800-1600	121	
12 but less than 20	250-525	250-450	120	0.5
		450-800	115	
		800-2100	111	

NOTE: The range of audibility in the table above is for information and is approximately the range at which a whistle may usually be heard on its forward axis in conditions of still air on board a vessel having average background noise level at the listening posts (taken to be 68 dB in the octave band centered on 250 HZ and 63 dB in the octave band centered on 500 HZ).

In practice the range at which a whistle may be heard is extremely variable and depends critically on weather conditions; the values given can be regarded as typical but under conditions of strong wind or high ambient noise level at the listening post the range may be much reduced.

86.07 Directional properties

The sound pressure level of a directional whistle shall be not more than 4 dB below the sound pressure level specified in section 86.05 in any direction in the horizontal plane within +/- 45 degrees of the forward axis. The sound pressure level of the whistle at any other direction in the horizontal plane shall not be more than 10 dB less than the sound pressure level specified for the forward axis, so that the range of audibility in any direction will be at least half the range required on the forward axis. The sound pressure level shall be measured in that one-third octave band which determines the audibility range.

86.09 Positioning of whistles
(a) When a directional whistle is to be used as the only whistle on the vessel and is permanently installed, it shall be installed with its forward axis directed forward.

(b) *Same as International Rules Annex III, Part 1(e), paragraph 2.*

86.11 Fitting of more than one whistle
If whistles are fitted at a distance apart of more than 100 meters, they shall not be sounded simultaneously.

86.13 Combined whistle systems
(a) A combined whistle system is a number of whistles (sound-emitting sources) operated together. For the purposes of the Rules a combined whistle system is to be regarded as a single whistle.

(b) The whistles of a combined system shall:
    (1) Be located at a distance apart of not more than 100 meters,
    (2) Be sounded simultaneously,
    (3) Each have a fundamental frequency different from those of the others of at least 10 Hz, and
    (4) Have a tonal characteristic appropriate for the length of vessel which shall be evidenced by at least two-thirds of the whistles in the combined system having fundamental frequencies falling within the limits prescribed in section 86.03, or if there are only two whistles in the combined system, by the higher fundamental frequency falling within the limits prescribed in section 86.03.
    NOTE: If due to the presence of obstructions the sound field of a single whistle or of one of the whistles referred to in section 86.11 is likely to have a zone of greatly reduced signal level, a combined whistle system should be fitted so as to overcome this reduction.

86.15 Towing vessel whistles.
A power driven vessel normally engaged in pushing ahead or towing alongside may, at all times, use a whistle whose characteristic falls within the limits prescribed by section 86.03 for the longest customary composite length of the vessel and its tow.

Subpart B — Bell or Gong

86.21 Intensity of signal
A bell or gong, or other device having similar sound characteristics shall produce a sound pressure level of not less than 110 dB at 1 meter.

86.23 Construction
Bells and gongs shall be made of corrosion-resistant material and designed to give a clear tone. The diameter of the mouth of the bell shall be not less than 300 mm for vessels of more than 20 meters in length, and shall be not less than 200 mm for vessels of 12 to 20 meters in length. The mass of the striker shall be not less than 3 percent of the mass of the bell. The striker shall be capable of manual operation. NOTE: When practicable, a power-driven bell striker is recommended to ensure constant force.

INLAND RULES ANNEX IV
Distress Signals

87.1 Need of assistance.
(a) through (o) same as International Rules Annex IV, Part 1, with this additional signal for Inland waters only:

(p) A high intensity white light flashing at regular intervals from 50 to 70 times per minute.

87.3 Exclusive Use
Same as International Rules Annex IV, Part 2.

87.5 Supplemental Signals
Same as International Rules Annex IV, Part 3, plus the additional reference to signals in International Telecommunication Union Regulations

INLAND RULES ANNEX V
Pilot Rules

88.01 Purpose and applicability
This part applies to all vessels operating on United States inland waters and to United States vessels operating on the Canadian waters of the Great Lakes to the extent there is no conflict with Canadian law.

88.03 Definitions
The terms used in this part have the same meaning as defined in the Inland Navigation Rules Act of 1980.

88.05 Copy of Rules
After January 1, 1983, the operator of each self-propelled vessel 12 meters or more in length shall carry on board and maintain for ready reference a copy of the Inland Navigational Rules.

88.09 Temporary exemption from light and shape requirements when operating under bridges.
A vessel's navigation lights and shapes may be lowered if necessary to pass under a bridge.

88.11 Law enforcement vessels
(a) Law enforcement vessels may display a flashing blue light when engaged in direct law enforcement or public safety activities. This light must be located so that it does not interfere with the visibility of the vessel's navigation lights.

(b) The blue light described in this section may be displayed by law enforcement vessels of the United States and the States and their political subdivisions.

88.12 Public Safety Activities
(a) Vessels engaged in government sanctioned public safety activities, and commercial vessels performing similar functions, may display an alternately flashing red and yellow light signal. This identification light signal must be located so that it does not interfere with the visibility of the vessel's navigation lights. The identification light signal may be used only as an identification signal and conveys no special privilege. Vessels using the identification light signal during public safety activities must abide by the Inland Navigation Rules, and must not presume that the light or the exigency gives them precedence or right of way.

(b) Public safety activities include but are not limited to patrolling marine parades, regattas, or special water celebrations; traffic control; salvage; firefighting; medical assistance; assisting disabled vessels; and search and rescue.

88.13 Lights on barges at bank or dock
(a) The following barges shall display at night and, if practicable, in periods of restricted visibility the lights described in paragraph (b) of this section:
    (1) Every barge projecting into a buoyed or restricted channel.
    (2) Every barge so moored that it reduces the available navigable width of any channel to less than 80 meters.
    (3) Barges moored in groups more than two barges wide or to a maximum width of over 25 meters.
    (4) Every barge not moored parallel to the bank or dock.

(b) Barges described in paragraph (a) shall carry two unobstructed white lights of an intensity to be visible for at least one mile on a clear dark night, and arranged as follows:
    (1) On a single moored barge, lights shall be placed on the two corners farthest from the bank or dock.
    (2) On barges moored in group formation, a light shall be placed on each of the upstream and downstream ends of the group, on the corners farthest from the bank or dock.

(3) Any barge in a group, projecting from the main body of the group toward the channel, shall be lighted as a single barge.

(c) Barges moored in any slip or slough which is used primarily for mooring purposes are exempt from the lighting requirements of this section.

(d) Barges moored in well-illuminated areas are exempt from the lighting requirements of this section. These areas are as follows:

Chicago Sanitary Ship Canal
(1) Mile 293.2 to 293.9
(3) Mile 295.2 to 296.1
(5) Mile 297.5 to 297.8
(7) Mile 298 to 298.2
(9) Mile 298.6 to 298.8
(11) Mile 299.3 to 299.4
(13) Mile 299.8 to 300.5
(15) Mile 303 to 303.2
(17) Mile 303.7 to 303.9
(19) Mile 305.7 to 305.8
(21) Mile 310.7 to 310.9
(23) Mile 311 to 311.2
(25) Mile 312.5 to 312.6
(27) Mile 313.8 to 314.2
(29) Mile 314.6
(31) Mile 314.8 to 315.3
(33) Mile 315.7 to 316
(35) Mile 316.8
(37) Mile 316.85 to 317.05
(39) Mile 317.5
(41) Mile 318.4 to 318.9

(43) Mile 318.7 to 318.8
(45) Mile 320 to 320.3
(47) Mile 320.6
(49) Mile 322.3 to 322.4
(51) Mile 322.8
(53) Mile 322.9 to 327.2
Calumet Sag Channel
(61) Mile 316.5
Little Calumet River
(71) Mile 321.2
(73) Mile 322.3
Calumet River
(81) Mile 328.5 to 328.7
(83) Mile 329.2 to 329.4
(85) Mile 330, west bank to 330.2
(87) Mile 331.4 to 331.6
(89) Mile 332.2 to 332.4
(91) Mile 332.6 to 332.8
Cumberland River
(101) Mile 126.8
(103) Mile 191

88.15 Lights on dredge pipelines

Dredge pipelines that are floating or supported on trestles shall display the following lights at night and in periods of restricted visibility.
(a) one row of yellow lights. The lights must be:
  (1) Flashing 50 to 70 times per minute,
  (2) Visible all around the horizon,
  (3) Visible for at least 2 miles on a clear dark night,
  (4) Not less than 1 and not more than 3.5 meters above the water,
  (5) Approximately equally spaced, and
  (6) Not more than 10 meters apart where the pipeline crosses a navigable channel. Where the pipeline does not cross a navigable channel the lights must be sufficient in number to clearly show the pipeline's length and course.

(b) Two red lights at each end of the pipeline, including the ends in a channel where the pipeline is separated to allow vessels to pass (whether open or close). The lights must be:

(1) Visible all around the horizon, and

(2) Visible for at least 2 miles on a clear dark night, and

(3) One meter apart in a vertical line with the lower light at the same height above water as the flashing yellow light.

INLAND INTERPRETATIVE RULES

AUTHORITY: 33 U.S.C. 2071, 49 CFR 1.46(n)(14).

90.1 Purpose.

(a) This part contains the interpretative rules for the Inland Rules. These interpretative rules are intended as a guide to assist the public and promote compliance with the Inland Rules.

90.3 Pushing vessel and vessel being pushed: Composite unit.

Rule 24(b) of the Inland Rules states that when a pushing vessel and a vessel being pushed ahead are rigidly connected in a composite unit, they are regarded as a power-driven vessel and must exhibit the lights prescribed in Rule 23. A "composite unit" in interpreted to be the combination of a pushing vessel and a vessel being pushed ahead that are rigidly connected by mechanical means so they react to sea and swell as one vessel. Mechanical means does not include lines, wires, hawsers or chains. [CGD 83-011, 48 FR 51621, Nov. 10, 1983]